India Studies in Business and Economics

The Indian economy is considered to be one of the fastest growing economies of the world with India amongst the most important G-20 economies. Ever since the Indian economy made its presence felt on the global platform, the research community is now even more interested in studying and analyzing what India has to offer. This series aims to bring forth the latest studies and research about India from the areas of economics, business, and management science. The titles featured in this series will present rigorous empirical research, often accompanied by policy recommendations, evoke and evaluate various aspects of the economy and the business and management landscape in India, with a special focus on India's relationship with the world in terms of business and trade.

More information about this series at http://www.springer.com/series/11234

Rajat Kathuria · Saon Ray
Kuntala Bandyopadhyay
Editors

Low Carbon Pathways for Growth in India

Springer

Editors
Rajat Kathuria
Indian Council for Research
on International Economic Relations
New Delhi
India

Kuntala Bandyopadhyay
Indian Council for Research
on International Economic Relations
New Delhi
India

Saon Ray
Indian Council for Research
on International Economic Relations
New Delhi
India

ISSN 2198-0012 ISSN 2198-0020 (electronic)
India Studies in Business and Economics
ISBN 978-981-13-0904-5 ISBN 978-981-13-0905-2 (eBook)
https://doi.org/10.1007/978-981-13-0905-2

Library of Congress Control Number: 2018943380

Printed on acid-free paper

This Springer imprint is published by the registered company Springer Nature Singapore Pte Ltd.
The registered company address is: 152 Beach Road, #21-01/04 Gateway East, Singapore 189721, Singapore

Acknowledgements

The genesis of this book lies in the project on Low Carbon Pathways for Growth in India which ICRIER undertook in 2014–15. The project was funded by the Foreign and Commonwealth Office, British High Commission; their support is gratefully acknowledged. The objective of the project was first to examine low carbon pathways and peaking years to help Indian policy makers take an informed decision and, second, to analyze the cost of action of mitigation versus the cost of inaction (climate impacts) for India. The third element was the examination of the key factors that influence existing models. The first paper, "Low Carbon Pathways," was written by Himanshu Gupta, the second paper, "Cost of Inaction on Mitigating Climate Change," by Dr. Vaibhav Chaturvedi of CEEW, and the third paper, "Assessment of Climate Models," was done at ICRIER.

As a part of the project, two workshops were held in New Delhi. The inception workshop, "Low Carbon Pathways for Growth in India," was held on the occasion of the visit of the UK Cabinet Minister Rt. Hon. Edward Davey, MP, Secretary of State for Energy and Climate Change, on July 25, 2014. The speakers at the workshop included Edward Davey, Shri Montek Singh Ahluwalia, Mrs. Naina Lal Kidwai, Dr. Navroz Dubash, and Dr. Sreekanth Gupta. A second stakeholder consultation on "Low Carbon Pathways for Growth in India" was organized on the occasion of the visit of Sir David King, on January 13, 2015, also in New Delhi. The findings of the study, including papers by Dr. Vaibhav Chaturvedi and Himanshu Gupta, were presented at these workshops. The speakers included Sir David King, Dr. Kirit Parikh, and Dr. Radhika Khosla. Both workshops were every well attended and generated a great deal of interest. Participants at both the workshops contributed to the discussion, and their comments are gratefully acknowledged.

Since the implementation of the low carbon pathways will occur through different sectors of the economy, subsequent to the third workshop on April 10, 2015, sectoral papers were conceived and written. The papers were reviewed, and the comments received have helped give the book its present shape. Dr. Vaibhav Chaturvedi and Dr. Prasoon Agarwal of Global Green Growth Institute commented on Himanshu Gupta's paper. Ms. Kanika Kalra of the Institute of Urban Transport commented on Zeba Aziz's paper, "Low Carbon Pathways—Opportunities in the Urban Transport Sector." Professor Meeta Keswani Mehra of Jawaharlal Nehru University and Prof. Surender Kumar of Delhi University commented on Dr. Saon Ray's and Dr. Nandini Kumar's paper "Air Quality and Industry." Dr. Neha Tripathi of the School of Planning and Architecture reviewed Dr. Indro Ray's paper, "Pay Less-get More: Energy Efficiency Approach to Urban Water Supply." Dr. Jenia Mukherjee of the Institute of Development Studies, Kolkata, commented on Dr. Nilanjan Ghosh's paper, "Water, Ecosystem Services and Food Security: Avoiding the Costs of Ignoring the Linkage." Two chapters, Indro Ray's "Sunny Side Up: India's Journey to Energy Security" and Subhomoy Bhattacharjee's "Energy Security Options for India in the Context of Great Power Rivalry Merging in the Indian Ocean," were added post the review of the manuscript at Springer. All the reviewers are gratefully acknowledged.

Finally, the guidance of Dr. Rajat Kathuria, Director and CEE, ICRIER, throughout the project and the steering committee of Prof. Partha Sen, Prof. Sreekanth Gupta, and Dr. Daljit Singh at the initial stages of conceptualization of the problem is also recognized.

The paper "Decarbonization of the Indian Railways" was part of the project 'De-carbonization of Indian Railways (Demand side)' conducted by ICRIER in 2015–16 and supported by New Climate Economy (NCE). The project was led by Dr. Saon Ray (ICRIER) and supported by Ms. Kuntala Bandyopadhyay (ICRIER) and Dr. T S Ramakrishan (T. A. Pai Management Institute). The technical work was reviewed by an advisory committee which includes Dr. Shuba Raghavan (Lumina Decision Systems), Mr. Stephen Perkins and Mr. Lorenzo Casullo (International Transport Forum, OECD), and Mr. Willy Bontinck (International Union of Railways and Belgian Railways), and benefitted from extensive stakeholder engagement from a number of experts in India including Mr. Sudhir Garg, (Indian Railways), Mr. Sudhir Saxena (REMCL), Dr. Rajat Kathuria (ICRIER) and Mr. Christoph Wolff (European Climate Foundation).

Contents

Editors and Contributors

About the Editors

Rajat Kathuria is Director and Chief Executive at Indian Council for Research on International Economic Relations (ICRIER), New Delhi. He has over 20 years of experience in teaching and 15 years of experience in economic policy, besides research interests on a range of issues relating to regulation and competition policy. He has taught undergraduate economics at the University of Maryland and is currently on leave from the International Management Institute (IMI), New Delhi, where he teaches managerial economics and international trade. He has worked with the World Bank, Washington DC, as Consultant and carried out project assignments for a number of international organizations, including ILO, UNCTAD, LIRNEasia, World Bank, and ADB. He has published in international and national journals, besides in popular magazines and newspapers. He is Founder Member of Broadband Society for Universal Access and served on the Board of Delhi Management Association. He currently is Independent Director on the Microfinance Institutions Network (MFIN) and on several government committees. He is also on the research advisory council of SBI. He has an undergraduate degree in Economics from St. Stephen's College, a master's from Delhi School of Economics, and a Ph.D. from the University of Maryland, College Park.

Saon Ray is Senior Fellow, Indian Council for Research on International Economic Relations (ICRIER), New Delhi. An economist specializing in the industry and international trade issues, her areas of interest include global value chains, technological upgrading of Indian industries, free trade agreements and trade creation effects, technology transfer, foreign direct investment, efficiency and productivity of firms, energy- and climate change-related issues. She has published widely on these issues in books and journal articles. Her latest book "Global Value Chains and the Missing Links: Cases from Indian Industry" was published by Taylor & Francis in 2018. Her PhD in Economics from the Jawaharlal Nehru University examined the role of intellectual property rights in transferring technology to developing countries.

Kuntala Bandyopadhyay is Research Associate in ICRIER. Her research interests are development economics, particularly in the areas of energy and climate issues, trade and labor issues. She is currently working in the New Climate Economy project which is looking into the feasibility of decarbonization of Indian Railways. She has an M.Phil. in Economics from Jawaharlal Nehru University and is currently pursuing her Ph.D. in the same university.

Contributors

Zeba Aziz is Consultant with ICRIER. Being an urban planner by training, she has worked on several urban realm projects in the past, the most recent being the Rockefeller Foundation-funded project on climate change and the economic competitiveness of cities where she focused on the urban development and transportation aspects. She is currently working on research projects under the Global Commission on New Climate Economy. Prior to joining ICRIER, Zeba worked as Town Planner with the City of Champaign, Illinois, during 2007–2011 where she worked on the issues of compact growth, infill development, green codes, and sustainable transportation options, among others. She also worked as Planning Consultant with IDFC in 2011–12 for planning of investment corridors in Madhya Pradesh, India, as well as with ILFS in 2012–13 for development of area improvement plans for the newly developed integrated freight complex at Ghazipur, Delhi. She received her master's in Urban and Regional Planning from the University of Illinois at Urbana Champaign and bachelor's in Physical Planning from the School of Planning and Architecture, New Delhi.

Subhomoy Bhattacharjee is Consulting Editor at the Business Standard newspaper. He works on public policy, primarily finance and energy. His latest book, "India's Coal Story," traces how India's coal reserves were at the centre of a major political scandal that nearly sent a prime minister to jail. It explores why since independence Indian business and government could not settle the rights on energy security, creating the murky politics of coal, and sketches the options for India's future energy security. His earlier book was "Special Economic Zones in India: Myths and Realities" (co-authored). He has read Economics at the Delhi School of Economics, Delhi University. He has worked in the Government of India and has worked with The Economic Times, The Indian Express and The Financial Express. Being a Consultant with Research and Information System for Developing Countries (RIS), he also often appears as Commentator on television channels on their business news programmes.

Vaibhav Chaturvedi is Research Fellow at the Council on Energy, Environment and Water (CEEW) and leads the Council's "Low Carbon Pathways" research. Prior to joining the Council, he was Postdoctoral Research Associate at the Joint Global Change Research Institute/ Pacific Northwest National Laboratory, USA. His research focuses on Indian and global energy and climate change mitigation policy issues within the integrated assessment modeling framework of the Global Change Assessment Model (GCAM). His recent work includes studies on pathways and policies for achieving India's nationally determined contributions (NDCs) and mid-century strategies within the context of sustainable development and national priorities, the climate policy—energy—water nexus, transportation energy and emission scenarios, hydrofluorocarbon emission scenarios and mitigation policy, and nuclear energy scenarios for India. He has been actively involved in global model comparison exercises like the Asian Modelling Exercise (AME) and the Energy Modelling Forum (EMF). He has been a part of Government of India committees for advising the Indian government on issues related to energy and climate policy. He actively publishes in, and reviews articles for, leading international energy and climate policy journals. He has a doctorate in economics from the Indian Institute of Management Ahmedabad and a master's in forest management from the Indian Institute of Forest Management, Bhopal.

Kaushik Deb is Research Fellow at KAPSARC focusing on natural gas markets. He previously led the analysis of the global natural gas markets and macroeconomic developments in the Asia-Pacific region in the economics team in BP. Among his earlier roles, in IDFC, his portfolio included policy research and advocacy on infrastructure and environmental economics issues. With a doctorate of sciences from the ETH Zurich, he has also guided and implemented research in applied economics in TERI University. He also was Program Director of the MBA programs at the University.

Nilanjan Ghosh is Senior Fellow and Head of Economics at Observer Research Foundation (ORF), Kolkata, and Chief Economic Advisor at the World Wide Fund for Nature (WWF) India, New Delhi. He has recently set up the Water Governance Studies division at ORF, Kolkata. His previous positions include stints as Chief Economist in the financial sector and Faculty at TERI University, New Delhi. He had been Visiting Fellow at the Linnaeus University, Sweden. A natural resource economist and econometrician by training, he obtained his PhD from the Indian Institute of Management (IIM) Calcutta and also visited Massachusetts Institute of Technology (MIT), USA, for certification in the Harvard University—MIT—Tufts University-housed program on water diplomacy. His research interests are ecological economics, trade and developmental issues, water resources, and financial markets. He has conducted consultancy/advisory assignments for a large number of organizations that include IUCN, Bangkok; FORMAS—The Swedish Research Council; Chinese Consulate, Kolkata; WWF UK; IWMI-WLE; World Bank; and many other clients. One of the foremost development analysts and ecological economists of the country, his publications include seven books and monographs (authored and/or edited) and more than 40 peer-reviewed research papers in journals and as chapters in edited volumes. He has been a regular columnist in Indian national dailies and portals like The Hindu Business Line, The Economic Times, The Third Pole, WION, ABP. He was Vice President and Acting President of the Indian Society for Ecological Economics (INSEE) during 2012–14, and delivered the presidential address at the Seventh Biennial Conference of INSEE in December 2013.

Himanshu Gupta is former Advisory Fellow to Vice President Al Gore and the Co-founder of ClimateAI, which uses AI to help CEOs, investment managers, and governments build climate resilience strategies. Prior to this, he was Project Leader of the India Energy Security Scenarios-2047 project for Government of India and co-authored the chapter on Renewable Energy for the 12th five-year plan with Montek Singh Ahluwalia. For his contributions to climate change space in India, he was included in the Forbes 30 under 30 list for India. He is a dual-degree graduate—MBA/MS in climate change from Stanford Universtiy—and an undergraduate in electrical engineering from IIT Kharagpur.

Manoj Kumar has around 18 years of experience which includes commercial and business development in oil and gas sector, fund-raising, economic and financial analysis and business planning in a variety of assignments. He is currently working with BP as Executive Assistant to Head of Country in India. Previously, he has worked with companies such as Cairn Energy, Reliance Industries, and Essar Oil. He has a very good understanding of upstream business. His qualifications include a degree in Chartered Financial Analyst (CFA) from the Institute of Chartered Financial Analyst of India and Postgraduate Diploma in Business Administration (PGDBA) specializing in Finance.

Nandini Kumar has been associated with the development of interdisciplinary programs in the environmental arena for about a decade at TERI University, New Delhi. Her doctorate is in atmospheric chemistry, and her area of interest is the application of systems thinking in environmental and sustainability issues, especially circular economies and industrial symbiosis. She won a Fulbright Environmental Leadership Fellowship in 2012 to MIT Sloan School of Management where she spent a semester getting acquainted with the System Dynamics modeling tool. She is currently Consultant at the CII-ITC Centre of Excellence for Sustainable Development, engaged in developing activities in the circular economy and sustainable public procurement.

T. S. Ramakrishnan has obtained his doctorate in Public Systems from Indian Institute of Management Ahmedabad, India, and is an independent consultant and researcher. His consulting, research, and teaching interests lie in public policy, transport systems with a focus on rail theme spanning across conventional rail, high-speed rail, metro and regional rail, integrated transport policy, public–private partnership, financing and structuring of infrastructure projects, public finance, and governance and democracy.

Indro Ray is Senior Analyst with Rio Tinto. He was Research Fellow at Indian Council for Research on International Economic Relations (ICRIER), New Delhi. His work looks at urban service delivery in the face of climate risks, and regional economic and urban growth trajectories. He was also the Technical Lead for the spatial analysis of rural prioritization in telecommunication. His research underpinnings on urban issues and climate change adaptation have been highlighted through his multiple publications and conference presentations. He has a bachelor's in Physical Planning from the School of Planning and Architecture, New Delhi, and a master's and a doctorate in Urban Planning from Arizona State University, USA.

Abbreviations

AA	Activity analysis
AEEI	Autonomous energy efficiency improvement
BAU	Business as usual
BC	Black carbon
BCM	Billion cubic meters
BEE	Bureau of Energy Efficiency
BIG	Baseline inclusive growth
BP	British Petroleum
BRTS	Bus rapid transport system
Btu	British thermal units
CAGR	Compound annual growth rate
CAP-EX	Capital expenditure
CCS	Carbon capture and storage
CEEW	Council of Energy, Environment and Water
CFA	Central Financial Assistance
CGE	Computable general equilibrium
CHP	Combined heat and power
CII	Confederation of Indian Industry
CO_2	Carbon dioxide
CO_2e	Equivalent carbon dioxide
COP	Conference of the Parties
CSE	Centre for Science and Environment
CT	Carbon tax
DICE	Dynamic Integrated Climate Economy
DISCOM	Distribution company
E&P	Exploration and production
EAI	Energy Alternatives India
EJ	Exajoule
EKC	Environmental Kuznets curve
EPRI	Electric Power Research Institute

ESCAP	Economic and Social Commission for Asia and the Pacific
ESMAP	Energy Sector Management Assistance Program
FAR	Fifth Assessment Report
FY	Financial year
GBI	Generation-based incentives
GCAM	Global Change Assessment Model
GDP	Gross domestic product
GEF	Global Environment Facility
GHG	Greenhouse gases
GIS	Geographic information system
GJ/t	Gigajoule/tonne
GoI	Government of India
GT	Giga tonnes
GW	Gigawatt
IAM	Integrated assessment model
ICRIER	Indian Council for Research on International Economic Relations
IEA	International Energy Outlook
IESS 2047	India Energy Security Scenarios 2047
IIP	Index of industrial production
IMF	International Monetary Fund
INDC	Intended Nationally Determined Contribution
INR	Indian Rupee
IPCC	Intergovernmental Panel on Climate Change
iPETS	Integrated Population Economy Technology Science
IPGPL	India Ports Global Private Limited
IRADe	Integrated Research for Action and Development
IRBM	Integrated River Basin Management
IREDA	Indian Renewable Energy Development Agency
IRWM	Integrated water resource management
JNNSM	Jawaharlal Nehru National Solar Mission
JNNURM	Jawaharlal Nehru National Urban Renewal Mission
Kb/d	Thousands of barrels per day
Kgoe	Kilogram(s) of oil equivalent
Kwh	Kilowatt-hours
$KWh/m^2/day$	Kilowatt-hours per square meter per day
LCIG	Low carbon inclusive growth
LNG	Liquefied natural gas
mb/d	Million barrels per day
MCA	Multi-criteria analysis
MLD	Million liters daily
MMTPA	Million metric tonnes per annum
MoEF	Ministry of Environment, Forest and Climate Change
MoWR	Ministry of Water Resources
MT	Metric tonnes
Mtoe	Million tons of oil equivalent

MU	Million units
Mwh	Megawatt hour
NAPCC	National Action Plan on Climate Change
NCAER	National Council of Applied Economic Research
NISE	National Institute of Solar Energy
NMT	Non-motorized mode of transportation
NPL	National Perspective Plan
NREL	National Renewable Energy Laboratory
NRW	Non-revenue water
NTFP	Non-timber forest products
NTPC	National Thermal Power Corporation
NUTP	National Urban Transport Policy
NVVN	National Vidyut Vyapar Nigam Limited
OECD	Organization of Economic Cooperation and Development
OM	Organic matter
p.a.	Per annum
PAT	Perform, achieve, and trade
PM	Particulate matter
ppm	Parts per million
ppmv	Parts per million by volume
PPP	Purchasing power parity
PUC	Pollution Under Control
PV	Photovoltaics
R&D	Research and development
RADAR	Radio detection and ranging
RESCO	Renewable Energy Service Company
RLP	River Link Project
RSPM	Respirable particulate matter
SAARC	South Asian Association for Regional Cooperation
SCADA	Supervisory control and data acquisition
SDC	Sagarmala Development Company
SECI	Solar Energy Corporation of India
SME	Small- and medium-sized enterprises
SS	Sustainable society
TAR	Third Assessment Report
tCO_2	Total carbon dioxide
TERI	The Energy and Resources Institute
TFPG	Total factor productivity growth
TPES	Total primary energy supply
TRACE	Tool for Rapid Assessment of City Energy
UDAY	Ujwal DISCOM Assurance Yojana
UK	United Kingdom
UN	United Nations
UNDP	United Nations Development Programme
UNEP	United Nations Environment Programme

UNFCCC	United Nations Framework Convention on Climate Change
UNIDO	United Nations Industrial Development Organization
US	United States
VMT	Vehicle miles traveled per capita
WEO	World Economic Outlook
WHO	World Health Organization

Part I
Low Carbon Pathways for India and Costs of Action Versus Inaction

Chapter 1
Overview

Rajat Kathuria, Saon Ray and Kuntala Bandyopadhyay

Abstract This chapter sets out the context for the book. The central question asked in this book is what are the low-carbon paths that India can adhere to without compromising its growth? This question has been answered in the various chapters that follow. The first part of the book examines the current energy demand and supply in India, the low-carbon pathways and the cost of inaction due to loss in yield of food crops such as maize, rice and wheat. The second part of the book provides a detailed discussion of the energy challenges faced in key sectors such as transport, industry and urban water supply. The chapter also provides an outline for the book.

1.1 Low-Carbon Pathways

The world's economy in 2014 grew at a rate of 2.5 percent, while India's grew at a rate of 6.8 percent. By 2030, it is expected that India will have a GDP of $ 10 trillion and become the largest economy after the US and China. By 2030, the world's population will reach close to 8 billion, with 85% residing in non-OECD countries. India's population currently is 1.28 billion and is expected to exceed 1.4 or 1.5 billion in 2030. India would become the most populous country in the world, overtaking China.[1]

[1] www.populationinstitute.org/resources/populationonline/issue/17/105/.

R. Kathuria · S. Ray (✉) · K. Bandyopadhyay
Indian Council for Research on International Economic Relations,
New Delhi, Delhi, India
e-mail: sray@icrier.res.in

R. Kathuria
e-mail: rkathuria@icrier.res.in

K. Bandyopadhyay
e-mail: kbandyopadhyay@icrier.res.in

R. Kathuria et al. (eds.), *Low Carbon Pathways for Growth in India*, India Studies in Business and Economics, https://doi.org/10.1007/978-981-13-0905-2_1

Population growth and economic growth are intrinsically linked to the energy demand of a country. Economic growth measures how an economy expands and is typically measured as the percentage increase in the gross domestic product in a year. The economic growth of a country depends on many factors, including productivity, demographic changes and energy consumption. Demographic factors influence economic growth of a country by changing the labour force participation.

Energy demand in India will increase drastically to support the population of 1.4 to 1.5 billion people in 2030. India's current energy demand is 788 Mtoe (in 2012, and including energy produced and net imports) and this is expected to rise to 1200 Mtoe in 2030. Due to the heavy use of fossil fuels, particularly coal, the challenge is to restrict its CO2 emissions and walk a low-carbon path.

1.2 Can India Grow in a Low-Carbon Way?

The central question that India needs to ask is: what are the ways in which a low-carbon path can be achieved without compromising growth? Objectives of alleviating energy poverty, continuing economic and social growth and development, and mitigating climate change may seem to stand in conflict with one another. For example, supplying an affordable and considerable amount of energy using low-cost energy fuels such as coal, potentially undermines efforts to mitigate climate change. Pursuing the use of domestic resources and promoting indigenous energy technologies instead of importing fuels could help enhance India's energy security in the long term, but does not solve its energy problems in the short term.

1.2.1 Energy Consumption and Projected Demand

India has three main energy policy objectives. First, providing access to energy, which is the foremost goal in energy policymaking. Nearly one-quarter of the population lacks access to electricity. Second, gaining energy security; dependence on imported fuels is increasing to meet the huge energy demand. This exposes the country to greater geopolitical risks and international price volatility. Finally, India is dedicated to mitigating the impacts of climate change. Despite being among the lowest emitters India is enthusiastic about its global responsibility: in its Intended Nationally Determined Contribution (INDC) India has said it will reduce emissions intensity by 30 to 35% by 2030 from the 2005 level.[2] In December 2009, India had announced that it would reduce the emissions intensity of its GDP by 20 to 25 percent, over the 2005 levels, by the year 2020. This forms the background for the book.

[2]http://www4.unfccc.int/submissions/INDC/Published%20Documents/India/1/INDIA%20INDC%20TO%20 UNFCCC.pdf.

1.3 An Outline of the Book

This book comprehensively overviews the low climate pathways open to India and their possible outcomes. It is organized in three parts: the first part reviews the current energy demand and supply scene in India, identifies the key factors that feed into existing models of climate change and also analyses the low-carbon pathways available to India, and the cost of action versus inaction. Kaushik Deb and Manoj Kumar's paper, India's *energy demand and supply,* presents a comprehensive picture of India's energy sector profile since 1980 analysing the movements of energy consumption and production profiles. Saon Ray and Kuntala Bandyopadhyay's paper, *Assessment of climate models*, critically assesses the available and widely used climate models, global and Indian, to measure the impacts of climate change. Himanshu Gupta's paper, *Low-carbon pathways*, discusses the low-carbon pathways open to India and finds that in the low-carbon scenario, emissions rise from 1.7 tonne/capita to 3.6 tonnes/capita by 2047, while India's energy demand rises by a factor of 3 from the base level of 420 Mtoe, in 2012, to MTOE in the year 2047. Vaibhav Chaturvedi's paper, *Cost of inaction on mitigating climate change*, estimates that the cost of inaction on GDP will be 0.45–1.19% in 2050, and 0.59–1.17% in 2100. The analysis includes the loss in yield of food crops such as maize, rice and wheat, the additional power required for cooling, and deaths due to diarrhoea, malaria and dengue.

The sectoral chapters in the second part of the book focus on more detailed analyses of the energy challenges and strategies in some key sectors. India is now at the cusp of an urban transition. Even though the country's urbanization has been low (32 percent) compared to that of countries such as China (45%) and Brazil (87%), this is set to change in the coming years with the urban population projected to almost double by 2030, adding another 200 million people to its urban areas.[3] With urbanization still nascent, there is a real opportunity to steer projected growth away from high carbon lock-ins and towards a low-carbon pathway. With growing urbanization, and with improving lifestyles and economic prosperity, the demand for resources such as water will be higher than ever. India is one of the most water-scarce countries in the world, and a high water demand together with unmanaged usage has already resulted in rapidly depleting water tables and drying of surface water sources. Indro Ray in his paper, *Pay less get more: energy efficiency approach to urban water supply*, discusses the energy efficiency approach to urban water supply. Transportation is a key contributor to carbon emissions on the national and urban scales. Zeba Aziz's paper, *Low-carbon pathways—opportunities in the urban transport sector*, focuses on the low-carbon strategies available to the urban transport sector. Industrial activity is the other major anthropogenic source of carbon dioxide. Saon Ray and Nandini Kumar's paper, *Air quality and industry,* discusses the ways to decarbonize industry in the context of India. Nilanjan Ghosh's paper, *Water, ecosystem services and food security: avoiding the costs of ignoring the*

[3]McKinsey Global Institute (2010).

linkage, challenges the linear and positive relation between water, critical ecosystem services and food security. It discusses the necessity of embracing a new and integrated paradigm of water management to embark on a low-carbon growth trajectory.

Indro Ray's paper, *Sunny Side Up: India's journey to energy security*, examines the solar targets set by the government as part of its INDC and argues that while significant amount of resources have been earmarked, implementation challenges remain. The challenges will have to be addressed if India wants to achieve its solar targets.

The concluding part of the book offers policy implications for low-carbon growth of India. Subhomoy Bhattacharjee's paper, *Energy security options for India in the context of great power rivalry emerging in the Indian Ocean*, points out the importance of the Indian Ocean for maritime and global sea trade. This is particularly due to three busiest shipping lanes between Asia and the world. A large part of the India's imports of natural gas occurs through the Indian Ocean which acts as a giant pipeline for both India and China. The paper argues the need for a sea borne energy focus as part of the integrated energy policy of India.

The final chapter, *Conclusions,* summarizes and concludes the book.

Reference

McKinsey Global Institute (2010). India's urban awakening: Building inclusive cities, sustaining economic growth

Chapter 2
India's Energy Demand and Supply

Kaushik Deb and Manoj Kumar

Abstract This chapter examines India's energy demand profiles and supply options. Unlike other markets, the energy sector is slow moving and changes in consumption and production profiles are a result of lumpy investment decisions as also gradual improvements in efficiency. As a result, India's energy sector profile appears unchanged since 1980, with coal and oil dominating the energy mix. This broad average, however, masks a significant shift away from coal and towards oil until 2000s, and the subsequent recovery in coal's share. The competition between coal and oil in the last century has now been played out between coal and gas over a much shorter period. The outcome has been an energy mix that is dominated by coal with some gains being made by renewables. In addition, India remains significantly import dependent for all forms of fossil fuels.

Unlike other markets, the energy sector is slow moving and changes in consumption and production profiles are a result of lumpy investment decisions as also gradual improvements in efficiency. As a result, India's energy sector profile appears unchanged since 1980, with coal and oil dominating the energy mix. This broad average, however, masks a significant shift away from coal and towards oil until 2000s, and the subsequent recovery in coal's share. The competition between coal and oil in the last century has now been played out between coal and gas over a much shorter period. The outcome has been an energy mix that is dominated by coal with some gains being made by renewables. In addition, India remains significantly import dependent for all forms of fossil fuels.

This scenario is unlikely to change into the future, with a sustained growth in population and GDP driving up energy consumption. This growth in demand would call for both higher volumes of production and imports. In addition, the large increase in demand calls for increased consumption of fossil fuels, especially coal. This is likely despite an expected rapid ramp up renewable energy and nuclear power generation, apart from gains in energy intensity.

K. Deb (✉) · M. Kumar
BP, Mumbai, India
e-mail: kaushikdeb@gmail.com

R. Kathuria et al. (eds.), *Low Carbon Pathways for Growth in India*, India Studies in Business and Economics, https://doi.org/10.1007/978-981-13-0905-2_2

2.1 Energy Sector Growth Since 1980

From a previous 3-decade average growth rate of 3.7%, the 1980s saw India's GDP growth rate rising much faster (DeLong 2003). In purchasing parity terms, this growth rate of 5.7% was significantly higher than the previous three decades since independence. In parallel, there was an increase in industrial activity faster than the increase in GDP during this period with a growth rate of 6.1% in manufacturing GDP. This resulted in rising primary energy consumption by 5.9% p.a. during the 1980s (BP 2015a). The growth was led by coal and oil, meeting the 49.5 and 33.6% of the growth in demand respectively. Domestic oil production, following the significant offshore find in Bombay High in 1974, rose by an average 6.3% p.a. during this decade, with oil and gas contributing more than a quarter of the increase in total production (26.6%). The vast majority of the increase in energy production though came from coal (70.8% of the total energy production during 1980–90). Interestingly, oil imports grew very slowly during this period with oil production keeping up with the growth in consumption, leading to a very gradual increase in oil imports by just 1% p.a. This is particularly remarkable given that the international price of oil fell during this period, and at its lowest was less than 40% of the decade's high.

2.2 Growth in the 1990s

After the sharp increase in GDP in 1980s, growth started to fall from a high of 8% in 1988 to just 2% in 1991. This deterioration has been attributed to the twin deficit —fiscal and current account, through the 1980s, and the reliance on external borrowings to bridge the gap (Panagariya 2004). The slowdown in GDP growth and in industrial production was mirrored in primary energy consumption. This slowdown mirrored a gradual maturity of the economy with a larger share of services and less energy intensive industry, as evidenced in a slower growth in manufacturing industry and IIP. The slowdown was more pronounced in coal and gas, reflecting the sharper decline in industrial activity. In particular, coal consumption growth, the primary driver in the early 1990s, stagnated in this period growing by only 4.2% during the 1990s (BP 2015a). On the other hand, a faster per capita growth in GDP sustained oil consumption growth at 6.2% p.a. As a result, oil matched coal as the lead contributor to energy demand. Overall, coal met 42.2% of the increased energy consumption during this period while the share of oil and gas combined rose to 53.1 from 46.1% during the 1980s. Among nonfossil fuels, nuclear overtook hydro as the principal fuel in terms of increments.

Production collapsed during the 1990s though (BP 2015a). A decade of under-investment as well as no incremental E&P in the 1990s led to oil production stagnating even as gas production continued to very slowly grow. The marginal increase in oil production in the first half of the 1990s (1.4% p.a. during 1990–95) was completely reversed by declines in the second half (−1.4% p.a. during 1995–2000).

Gas production growth also slowed down from a high of 26.2% p.a. in the 1980s to just 8.1% p.a. in the 1990s. Coal production growth also slowed down to 4.1% p.a. Overall, energy production increased by only 3.3% over the entire decade compared to 5% p.a. growth in energy consumption. The more rapid growth in energy consumption compared to production, particularly for oil, was supported by a weak international oil market that allowed oil imports to rise. Oil imports tripled during this decade from less than 500 Kb/d in 1990 to 1.54 mb/d in 2000, while Brent oil prices rose by just 20%. Coal imports also tripled during that decade though remained at modest levels; the share of domestic coal consumption met by imports increased from under 4–8.3%.

2.3 Recovery Between 2000 and 2010

The recovery in GDP growth rates during the first decade of the twenty-first century, and rising private investments in infrastructure were harbingers of a coming growth spurt. GDP and investment rose rapidly in this period, and this called for higher energy consumption. Even as the global economy spiraled into the worst financial crisis since the great depression in 2008 Reuters. (2009, February 29), India seemed to buck the trend. A previous 5-year GDP growth rate of 6.5% p.a. (over 2000–05) compared to the world average of 3.9% p.a., further increased to 8.3% p.a. during 2005–10, more than double the world average of 3.8% p.a., and well over the rest of the non-OECD (6.8% p.a.). India's energy consumption increased by 5.6% p.a. during 2000–10. This period was characterized by particularly strong growth in gas consumption, and gas met an even larger share of growth in consumption. Gas consumption grew by 9.1% p.a. meeting 15.3% of the increased in consumption, up from under 11.2% earlier in the century (BP 2015a). This was at the expense of oil, which despite stronger growth at 3.9%, lost share in the increment in consumption falling to 23% from 41.9% during the 1990s. Coal consumption growth also picked up to 6.1% p.a. This growth in energy consumption came about from a rapid increase in production with a growth rate of 5% p.a. compared to just 3.3% during 1990s.

2.4 Declines Since 2010

However, a surprising policy paralysis affected the economy in general, and the energy sector in particular from 2010 onwards. New E&P activity was stalled with a variety of unresolved issues related to licensing, price regulation, and other regulatory and administrative issues (Directorate General of Hydrocarbons 2014). In 2014, India's production was just 2.8% of the world's total (BP 2015a). This slowdown is attributable to gas and oil, which comprised 25.4% of total energy production in 2010 falling to 19.3% in 2014. Oil production growth fell to just 0.4% p.a. during this period while gas production started to decline sharply by 11.2% p.a. Coal production kept growing by 2.9%, even though lower than earlier. In fact, coal production in India fell twice in these 5 years, the first volumetric declines since 1999. Even so, coal regained back its share in energy production. This decline in energy production has had significant effects. Gas consumption declined (−5.2% p.a.), though more slowly than production, with coal consumption rising sharply (8.5% p.a.) to fill this space. However, since the increase in coal production was significantly lower than the increase in coal consumption, coal imports increased sharply to nearly twice of the 2010 levels. This has allowed electricity generation to continue rising at 7% p.a. even with coal and gas production not keeping up (Central Electricity Authority 2015). There have been increases in gas imports as well by 12.4% p.a. with India emerging in the top 5 LNG importers in the world. In general, coal consumption and imports have responded quickly to changes in gas availability, thus balancing the energy markets. Oil consumption growth also slowed as the commodity super cycle intensified. This reversal took India back to the 1980s in terms of fuel shares.

2.5 Implications

Over the last few years, even with economic growth slowing down in India, energy consumption has been robust. GDP growth slowed down from 7.4% p.a. during 2000–10 to 6.1% p.a. in 2010–14, while energy consumption growth is up from 5.6 to 5.8% p.a. As a result, improvements in energy intensity of GDP also slowed down. In addition, the evolving fuel mix had implications for CO_2 emissions from energy use as well. From a decline of 1.6% p.a. during 2000–10, energy intensity fell by 0.3% p.a. during 2010–14 (BP 2015a). In line with improvements in energy intensity, the growth CO_2 emissions from energy consumption continued to decline in India–from 5.6% p.a. during 2000–10 to 5.8% p.a. during 2010–14.

More significantly, the sharper slowdown in domestic production compared to consumption implied that the share of India's energy consumption met by domestic sources fell to 57% by 2014, the lowest on record (BP, 2015a). India's net energy imports[1] increased by 5.5% p.a. during 2010–14, compared to 4.8% in the first decade of the century. While coal led the trend with imports rising by 28.6% p.a. during this period, gas and oil imports rose by 12.4% p.a. and 4.9% p.a., respectively. Underlying this rapid increase in coal imports is the trend in domestic gas production during this period described above and the tight Asian LNG market. As domestic gas production collapsed in 2010, energy demand shifted to imports. LNG, on the other hand, entered a 3-year supply growth lull in 2011 along with the Fukushima nuclear disaster pushing Asian demand (and prices), to record highs and making gas imports much more expensive than coal imports. The result was this dramatic increase in coal imports during 2010–14.

[1]Net imports here are defined as the difference between consumption and production, ignoring stockbuilding and losses in transit.

2.6 Energy Demand Going Forward

Despite the growth in energy consumption, India's primary energy consumption per capita remains low compared to the world's average. Industrialization and economic growth are key factors that would influence the growth in India's energy demand as also the energy mix. The forecast presented below, based on the BP Energy Outlook, is a trajectory of the energy system, based on likely economic and population growth, as well as developments in policy and technology (BP 2015b). The GDP numbers are expressed in real 2011 US dollars and Purchasing Power Parity (PPP) exchange rates. Our population series are taken directly from the United Nations Population Division, United Nations (2015). Population growth and increases in income per person are the key drivers behind growing demand for energy. By 2035, the India's population exceeds 1.5 billion, which means an additional 250 million people will need energy. Over the same period, GDP is expected to treble, with India contributing nearly 13% of the total world's GDP growth.

2.6.1 *Demand*

The energy consumption in India is the fourth biggest after China, USA and Russia. Due to rapid economic expansion, India is one of the world's fastest growing energy markets and is expected to be one of the largest contributor to the increase in global energy demand by 2035, accounting for 16% of the rise in global energy demand (BP 2015b). India's demand is expected to increase by 128% from current 595–1355 Mtoe by 2035 (average per annum growth rate of 3.8%), with India accounting for 8% of the world's energy demand by 2035. During the same period, China will account for 26% of world's energy demand.

India has the world's 4th largest coal reserves and coal remains the largest primary energy contributor. Still, its share is likely to reduce from current 55% today to 50% by 2035, displaced replaced by cleaner fuels like nuclear and renewables. Renewables lead the nonfossil fuels, increasing their share in the energy mix to 6%. In comparison, the global share of coal is 26% by the year 2035. Oil (28%) will be the second largest fuel, with natural gas (8%) and nonfossil fuels (nuclear, hydro, and renewables) at 13% far behind by year 2035. Oil, gas, and coal demand is projected to grow in the range of average 3–5% per annum. Natural gas will account for 26% share in world's energy demand as against just 8% in India. Nuclear energy is expected to grow around four times from current 7.5–35 Mtoe in 2035 (7.2% p.a.). Insofar as renewables consumption, it is projected to increase at an average 9% per annum from current 12–80 Mtoe in 2035 led by solar (27% p.a.) and biofuels (13% p.a).

Industrial demand will account of 38% of overall energy demand by end 2035, growing at an average annual rate of 3.9% (BP 2015b). Share of coal in the overall industrial demand is projected to be 37% followed by oil (21.1%) and gas (13.6%)

by the year 2035. Transport sector will account for 17% of overall energy demand by end 2035 (average growth rate of 5.1%). Oil will dominate this sector with 94% share growing at an average 5.1%. It is estimated that power sector will account for 31% of the total energy demand by 2035. Energy consumption in power generation doubles (+124%) and while coal remains the dominant fuel source, its market share drops from 76% today to 72% in 2035. Given India's growing energy demands and limited domestic fossil fuel reserves, the country has ambitious plans to expand its renewable and nuclear power industries. Share of nuclear in power generation is expected to double from current 2.5–5% by 2035. Renewables creep up to 11% of the power mix, with hydro providing another 9% of the energy consumed for power generation.

Thus, there is no significant change envisaged in India's energy mix from current period to 2035. India's energy mix evolves very slowly over the next 22 years with fossil fuels accounting for 87% of demand in 2035, compared to a global average of 81%. This is down from 92% today. The fuel mix is dominated by coal, accounting for half of energy demand.

2.6.2 *Meeting Demand*

India's energy production is projected to keep pace with rising demand growing from current 350–780 Mtoe by 2035 (average growth rate of 3.7% per annum) (BP 2015b). Coal will continue its share of around 65% in total energy production of India throughout the period up to 2035 increasing from current 228–519 Mtoe by 2035. However, gap between demand and supply will continue to persist at around 40%. Share of oil in India's energy production is projected to reduce from current 12–4 % by 2035 with production falling from current 42–33 Mtoe by 2035. Imports will account for 90% of oil demand. For natural gas, the domestic production is expected to increase from current 30–54 Mtoe by 2035. This would only meet 50% of the natural gas demand by the year 2035. Renewables production is likely to keep pace with its demand growing at an average 9% per annum from current 12–80 Mtoe 2035. The share of renewables is expected to increase from current 3–10% by 2035. Nuclear energy is also expected to increase (7% per annum) from 7.5–35 Mtoe by 2035.

It is projected that by 2035 India remains import dependent despite increases in nonfossil fuel production. Oil imports will rise by 161% and account for 61% of the net increase in imports, followed in volumetric terms by increasing imports of coal (+96%) and gas (+270%).

Currently, India's net imports are nearly 135 Mtoe of crude oil, 16 Mtoe of LNG, and 95 Mtoe coal totalling to around 250 Mtoe of primary energy which is equal to around 40% of total primary energy demand.

References

BP (2015a) BP Energy Outlook 2035. BP, London

BP (2015b) Statistical Review of World Energy 2015. BP, London

Central Electricity Authority (2015) Growth of Electicity Sector in India from 1947–2015. Central Electricity Authority, Government of India, New Delhi

DeLong JB (2003) India since independence: an analytic growth narrative. In: Rodrik D (ed) In search of prosperity: analytic narratives on economic growth. Princeton University Press, New Jersey, pp 184–204

Directorate General of Hydrocarbons (2014) NELP-X oil & gas blocks: salient features. In: Petrotech 2014. Ministry of Petroleum & Natural Gas, Government of India, Noida

NITI Aayog (2017) Draft National Energy Policy of the NITI Aayog, Government of India. Aavailable at http://niti.gov.in/writereaddata/files/new_initiatives/NEP-ID_27.06.2017.pdf

Oxford Economics (2015) Global Economic Databank. Oxford Economics

Panagariya A (2004) India in the 1980s and 1990s: a triumph of reforms. (W. I. Paper, Ed.)

Reuters. (2009, February 29). Two top economists agree 2009 worst financial crisis since great depression; risks increase if right steps are not taken. Retrieved 1 20, 2015, from Reuters:http://www.reuters.com/article/2009/02/27/idUS193520+27-Feb-2009+BW20090227

United Nations (2015) World Urbanization Prospects: The 2014 Revisions. Population Database, Population Division, Department of Economic and Social Affairs, New York, United Nations

Chapter 3
Low Carbon Pathways for Growth in India: Assessment of Climate Models

Saon Ray and Kuntala Bandyopadhyay

Abstract Modelling the monetary impacts of climate change globally requires quantitative analysis of a very broad range of environmental, economic and social issues. Integrated Assessment Models (IAMs) provide a useful tool in this regard. Their estimates provide an important foundation for later work, and their results are valuable for informing policy. This chapter provides an overview of the existing models including the Mendelsohn, Dietz and Stern models. In addition it reviews the Indian models which include the NCEAR, TERI, IRADe and the McKinsey model. It also discusses the co-benefits approach proposed by Dubash (Econ Polit Weekly 48(22):47–62, 2013) in the Indian context.

3.1 Introduction

The world energy consumption is expected to grow by 56% between 2010 and 2040. Much of the growth in the energy consumption is expected from countries such as China and India, and will be driven by strong, long-term economic growth. For the past two decades, both these countries have been among the world's fastest growing economies; they have led the economic recovery from the recession has been led by these countries. Since 1990, their combined energy consumption accounted for 10% of the total world energy consumption in 1990, and 24% in 2010.

This paper reviews the key factors that feed into existing models of climate change.

S. Ray (✉) · K. Bandyopadhyay
Indian Council for Research on International Economic Relations,
New Delhi, Delhi, India
e-mail: sray@icrier.res.in

K. Bandyopadhyay
e-mail: kbandyopadhyay@icrier.res.in

R. Kathuria et al. (eds.), *Low Carbon Pathways for Growth in India*, India Studies in Business and Economics, https://doi.org/10.1007/978-981-13-0905-2_3

3.2 Review of the Models

Modeling the monetary impacts of climate change globally is very challenging: it requires quantitative analysis of a very broad range of environmental, economic and social issues. Integrated Assessment Models (IAMs) provide a useful tool in this regard. Their estimates provide an important foundation for later work, and their results are valuable for informing policy. However, these models are limited to snapshots of climate change at temperatures now likely to be exceeded by the end of this century. Below are three important examples of models of this category.

The Mendelsohn model (Mendelsohn et al. 2000) estimates impacts only for five "market" sectors: agriculture, forestry, energy, water, and coastal zones. The Tol model (Tol 2002) estimates impacts for a wider range of market and nonmarket sectors: agriculture, forestry, water, energy, coastal zones, and ecosystems, as well as mortality from vector-borne diseases, heat stress, and cold stress. The Nordhaus model (Nordhaus and Boyer 2000) includes a range of market and nonmarket impact sectors: agriculture, forestry, energy, water, construction, fisheries, outdoor recreation, coastal zones, mortality from climate-related diseases and pollution, and ecosystems. It also includes, what were at that time, pioneering estimates of the economic cost of catastrophic climate impacts.

Most formal models use 2–3 °C warming as a starting point. In this temperature range, the cost of climate change could be equivalent to a loss of 0–3% in global GDP from what could have been achieved in a world without climate change. Models differ on whether low levels of global warming would have positive or negative global effects. But all agree that the effects of warming above 2–3 °C would reduce global welfare, and that even mild warming would harm poor countries. Their results depend on key modeling decisions, including how each model values the costs to poor regions and what it assumed about societies' ability to reduce costs by adapting to climate change.

The existing estimates of monetary costs of climate change also omit significant factors such as such as extreme weather events, social and political instability, and cross-sectoral impacts.

Stern (2008) points out that business-as-usual (BAU) temperature increases may exceed 2–3 °C by the end of this century. Using an Integrated Assessment Model, and with due caution about the ability to model, he estimated the total cost of BAU climate change to equate to an average reduction in global per capita consumption of 5%, at a minimum, now and forever. Stern uses the PAGE2002 IAM (Hope 2006), which can take account of the range of risks by allowing outcomes to vary probabilistically across many model runs, with the probabilities calibrated to the latest scientific quantitative evidence on particular risks. He runs the model under two different assumed levels of climatic response. The "baseline climate" scenario is designed to give outputs consistent with the IPCC's Third Assessment Report (TAR) (IPCC 2001). The "high climate" scenario adds the risk of amplifying natural feedbacks in the climate system. Preliminary estimates of average losses in global per capita GDP in 2200 range from 5.3 to 13.8%, depending on the size of climate system feedbacks and what estimates of "nonmarket impacts" are included. In all scenarios, the highest impacts are in Africa and the Middle East, and India and Southeast Asia. In all scenarios, the consequences of climate change become disproportionately more severe with increased warming. Stern finds that the welfare costs of BAU climate change are very high. Climate change is projected to reduce average global welfare by an amount equivalent to a permanent cut in per capita consumption of a minimum of 5%. The reductions are larger if nonmarket impacts, feedbacks and regional costs are included. Putting these three factors together would probably increase the cost of climate change to the equivalent of a 20% cut in per capita consumption, now and forever.

Dietz and Stern (2014) assessed the series of Dynamic Integrated Climate-Economy (DICE) models (Nordhaus 1991). These models have inbuilt assumptions on growth, damages, and risk, which together result in gross underassessment of the overall scale of the risks from unmanaged climate change. The authors show that if the analysis is extended to take into account three essential elements of the climate problem—the endogeneity of growth, the convexity of damages, and climate risk—optimal policy comprises stronger controls. With the extended models, BAU trajectories of greenhouse gas emissions give rise to potentially large impacts on growth and prosperity in the future, especially after 2100. These impacts are large enough to feed back into future emissions via reduced activity, but the feedback is too small and too late for the system to self-regulate. As a guide, the authors find that the extended DICE models suggest the carbon price in a setting of globally coordinated policy, such as a cap-and-trade regime or a system of harmonized domestic carbon taxes, should be in the range \$32–103/tCO_2 (2012 prices) in 2015.

3.3 Results from the Indian Modeling Exercises[1]

India is one of the lowest emitters of greenhouse gases (GHGs) in the world on a per capita basis. At 1.4 tCO_2/person in 2010, India's emissions were less than one-third of the world average of 4.5 tCO_2/person, less than one-fourth that of China's, and one-twelfth that of the US's. But India is still threatened by the impact of global warming and climate change. Enthusiastic about its global responsibility, in December 2009, it announced that it would reduce the emissions intensity of its GDP by 20–25%, over the 2005 levels, by the year 2020. India is resolute about ensuring sustainable growth based on low carbon principles.

But this is not an easy task. The Low Carbon Society Vision 2050 India (2009) states that India faces challenges in economic development which have to be met with limited resources, minimal externalities, and in the presence of large uncertainties with respect to climate.

[1]Appendix 3.1 A–3.4 A lists the various assumptions and results obtained from the various models.

The Expert Group on Low Carbon Strategies for Inclusive Growth (2014) has evolved a macro-model to fully elucidate the inter-sectoral implications of different mitigation measures and ensure that the low carbon strategies being recommended are mutually consistent.

The model's output is summarized in two end-point scenarios: the BIG (Baseline, Inclusive Growth), and the LCIG (Low Carbon, Inclusive Growth). While inclusive actions remain unchanged between the two scenarios, low carbon strategies span the vector space between them. Pursuit of Low Carbon Strategies brings down the average GDP growth rate by 0.15 percentage points, while per capita CO_2 emissions (in 2030) fall from 3.6 tons in the BIG scenario to 2.6 tons in the LCIG scenario. However, in both scenarios, the total carbon emissions continue to rise up to the year 2030.

The cumulative costs of low carbon strategies have been estimated to be 834 billion US dollars at 2011 prices, over the two decades between 2010 and 2030. While total power demand remains unchanged between the two scenarios, emission intensity of GDP declines by 22%, over 2007 levels (by 2030) in the BIG scenario, as compared to 42%, over 2007 levels (by 2030) in the LCIG scenario. Further, due to a massive change in the energy mix by 2030, demand for coal comes down from 1568 Mt in the BIG to 1278 Mt in the LCIG scenario, demand for crude oil comes down from 406 Mt in the BIG to 330 Mt in the LCIG scenario, while demand for gas marginally rises from 187 bcm in the BIG to 208 bcm in the LCIG scenario. At the same time, the installed wind and solar power capacities need to be increased to 118 and 110 GW respectively, by the year 2030, in the LCIG scenario.

Low Carbon Society Vision 2050 India (2009) assesses two paradigms for transiting to low carbon future in India. The first pathway assumes a conventional development pattern together with a carbon price that aligns India's emissions to an optimal 450 ppmv CO_2e stabilization global response. The second emissions pathway assumes an underlying sustainable development pattern characterized by diverse response measures typical of "sustainability" paradigm. An integrated modeling framework is used for delineating and assessing the alternate development pathways having equal cumulative CO_2 emissions during the first half of the twenty-first century.

The base case scenario assumes future economic development along a conventional path. In the case of a developing country, such as India, the scenario assumes that future socioeconomic development will mimic the resource-intensive development path followed by the developed countries. Two paths are considered as part of Low Carbon Scenarios: the Conventional Path, Carbon Tax (CT) Scenario, and the Sustainable Society (SS) Scenario.

The CT scenario presumes a stringent carbon tax (or permit price) trajectory compared to a milder carbon regime assumed under the base case while sustainability is the rationale for the ST scenario. The perspective is a long-term one, aiming to deliver intergenerational justice by decoupling economic growth from the highly resource-intensive and environmentally unsound conventional path. In the SS scenario, mitigation choices are more diverse and include measures that are designed to influence several development indicators simultaneously. It pays greater attention to public investment decisions, such as infrastructure which lead to modal shifts in the transport sector; and institutional interventions that alter the

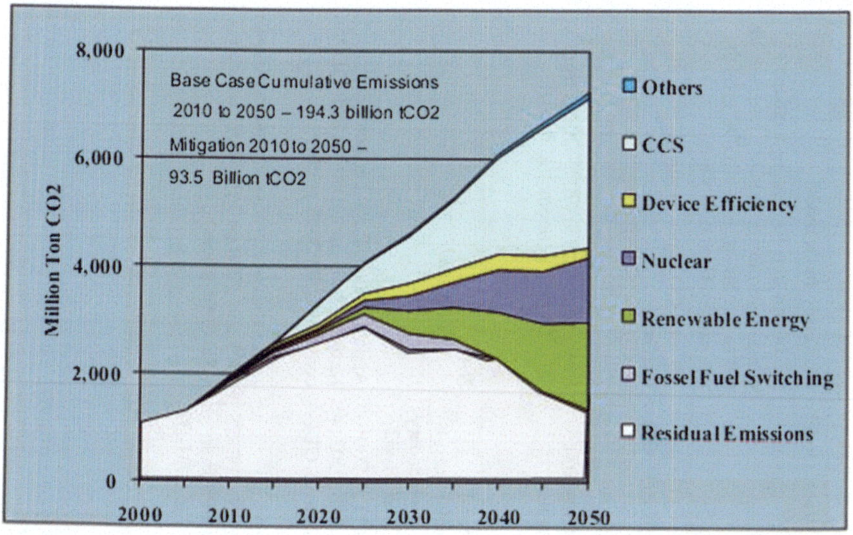

Fig. 3.1 Mitigation options in carbon tax scenario. *Source* Low carbon society vision 2050 India (2009)

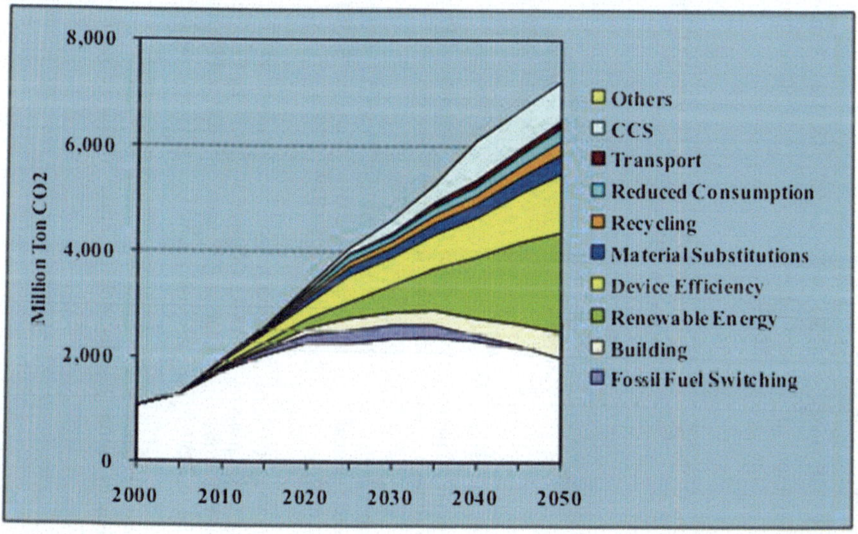

Fig. 3.2 Mitigations options in sustainability scenario. *Source* Low carbon society vision 2050 India (2009)

quality of development. In the CT scenario, the mitigation measures are more direct and have a greater influence on private investments (Figs. 3.1 and 3.2).

For realizing the vision of a Low Carbon Society for India, the study listed policy actions required to implement mitigation measures such as sustainable transport, low carbon electricity, fuel switching, building design, material

Table 3.1 Results for Illustrative Scenarios[a]

	NCAER CGE model	TERI MoEF model	IRADe AA model	TERI Poznan model	McKinsey India model
GHG emissions in 2030–31 (CO_2 or CO2e) (billion tons)	4.00 billion tons of CO2e	4.9 billion tons (in 2031–32)	4.23 billion tons	7.3 billion tons in 2031–32	5.7 billion tons (including methane emissions from agriculture); ranges from 5.0 to 6.5 billion tons if GDP growth rate ranges from 6 to 9%
Per capita GHG emissions in 2030–31 (CO_2 or CO2e)	2.77 tons CO2e per capita	3.4 tons CO2e per capita (in 2031–32)	2.9 tons CO2e per capita	5.0 tons CO2e per capita (in 2031–32)	3.9 tons CO2e per capita (2030), all GHGs
CAGR of GDP till 2030–2031, %	8.84%	8.84% (Exogenous–taken from CGE)	7.66% (Endogenous, 2010–11 to 2030–31)	8.2% 2030–2031 (Exogenous)	Exogenous–7.51% (2005–2030) from MGI Oxford Econometric Model
Commercial energy use in 2030-31, mtoe	1087 (Total commercial primary energy forms)	1567 (Total commercial Energy including secondary forms) in 2031–32	1042 (Total commercial Primary energy)	2149 (Total commercial energy including secondary forms) in 2031–32	NA
Fall in energy intensity	3.85% per annum (compound annual decline rate)	From 0.11 in 2001–02 to 0.06 in 2031–32 kgoe per $ GDP at PPP	From 0.1 to 0.04 kgoe per $ GDP at PPP	From 0.11 in 2001–02 to 0.08 in 2031–32 kgoe per $ GDP at PPP	Approximately 2.3% per annum between 2005 and 2030 (at PPP GDP, constant USD 2005 prices)
Fall in CO_2 (or CO2e) intensity	From 0.37 kg CO_2e to 0.15 kg CO2e per $ GDP at PPP from 2003–04 to 2030–31	From 0.37 to 0.18 kg CO_2 per $ GDP at PPP from 2001–02 to 2031–32	From 0.37 to 0.18 kg CO_2 per $ GDP at PPP from 2003–04 to 2030–31	From 0.37 to 0.28 kg CO_2 PER $ GDP at PPP from 2001–02 to 2031–32	Approximately 2% per annum between 2005 and 2030 (at PPP GDP, constant USD 2005 prices)

[a]Tables prepared by Gayatri Khedhkar

Source India's GHG Emissions Profile: Results of Five Climate Modelling Studies

Table 3.2 Assumptions and data sources for Illustrative Scenarios

	NCAER CGE model	TERI McEF model	IRADe AA model	TERI Poznam model	McKinsey India model
Assumptions	TFPG = 3.0% AEEI = 1.5% No new GHG mitigation policy	TFPG = 3.0% Energy Efficiency improvement consistent with AEEI assumptions in corresponding CGE run but constrained by limits to energy efficiency improvements in specific technologies as given in international published literature. No new GHG mitigation policy; discount rate = 15% financial costs	TFPG = 3.0% AEEI = 1.5% (amounting to 36.5% improvement in specific energy consumption from 2003 to 2030). No new GHG mitigation policy; max. savings rate = 3.5% social discount rate = 10% govt. Annual consumption increase = 9%	Efficiency improvements as per past trend and as per expert opinion considering level of maturity of specific technology in India. Discount rate = 10% economic costs, no new GHG mitigation policy	Sector by sector assumptions of demand and technology mix leading to illustrative scenario emissions
Data sources					
Population	Registrar General of India (till 2026, extrapolated at same rates till 2030)	Registrar General of India (till 2026, extrapolated at same rates till 2030)	Registrar\General of India (till 2026, extrapolated at same rates till 2030)	Registrar General of India (till 2026, extrapolated at same rates till 2030)	Registrar General of India (till 2026, extrapolated at same rates till 2030)
Global/ domestic energy price projections	International Energy Agency (WEO 2007) for international, endogenous for domestic	International Energy Agency (WEO 2007) for international; price indices from CGE model for domestic fuel prices; taxes and subsidies included to compute financial prices	International Energy Agency (WEO 2007) for international; endogenous for domestic	TERI estimates for both international and domestic prices based on prevailing market conditions	International Energy Agency for International energy prices

(continued)

Table 3.2 (continued)

	NCAER CGE model	TERI MoEF model	IRADe AA model	TERI Poznam model	McKinsey India model
GDP growth rates	Endogenous	Exogenous—from CGE output	Endogenous	Exogenous —8.2% (2001–2031)	Exogenous—7.51% (2005–2030) from MGI Oxford Econometric model
Foreign savings projections	Study by NCAER (2009)	NA	Endogenous	NA	NA
Domestic savings rate	National Accounts Statistics	NA	Max 35%	NA	NA
Specific Energy Technologies Data	NA	Data set of > 300 technologies compiled by TERI in study for Principal Scientific Adviser, and technology diffusion consistent with AEEI assumptions as reflected in CGE model	Eight electricity generation technologies (thermal, hydro, natural gas, wind, solar, nuclear, diesel, wood and more efficient coal technology)	Data set of > 300 technologies compiled by TERI in study for Principal Scientific Adviser with recent update	Data set of 200 Technologies incorporated in the McKinsey Global Cost Curve model, adapted for Indian values, capex and cost
GHG emissions coefficients	National Communications	National Communications	National Communications	National Communications	National Communications + IPCC (2001) + own estimates for power sector
Various other key parameters	Published Literature, NCAER and Jadavpur University estimates	Govt. of India Data, other published literature	Govt. of India Data	Govt. of India Data, own estimates, expert opinion, published literature	Govt. of India Data, own estimates

Source India's GHG Emissions Profile: Results of Five Climate Modelling Studies

Table 3.3 Shukla-AIM model comparison with MoEF models (Shukla et al. 2015)

GHG emissions in 2050	7.2 billion tons
Per capita emissions in 2050	4.5 tons
CAGR of GDP till 2032	8%
Commercial energy use in 2050, mtoe	2825 mtoe
Fall in energy intensity	3% per annum
Fall in CO_2 intensity	
Assumptions	
Population	UN population medium scenario version 2004
Global/domestic energy price projections	Global prices by IEA
GDP growth rates	Exogenous
Foreign savings projections	
Specific energy technologies data	Technologies in power, transport
Model/methodology descriptions	
Model/methodology type	AIM CGE/GCAM model
Key features of model/ methodology	Top down—bottom up integrated model soft linking of AIM CGE model with ANSWER- MARKAL model
Key inputs	Population, energy prices, GDP growth rate,
Key outputs	CO_2 emissions, energy intensity, CO_2 intensity, energy demand, mitigation choices
Number of sectors	13 sectors, industry divided into 11 subsectors
Greenhouse gases included	CO_2 (energy and industry only)
Primary energy forms	Coal, oil, gas, nuclear, hydro, biomass, renewable

Source India's GHG Emissions Profile: Results of Five Climate Modelling Studies

Table 3.4 Difference between IRADe–MoEF model and expert group macro-model

Parameter	IRADe MoEF model	Expert group macro-model
Results	Results till 2030 with 3-year interval	Results till 2030 for each year
TFPG assumptions	3%	1%- agriculture, 1.5%—non-agriculture sector
AEEI	1.5%	0.5% for BAU
Development indicators	No	Basic development indicators included which makes BAU scenario a BIG—baseline inclusive growth scenario

Source: India's GHG Emissions Profile: Results of Five Climate Modelling Studies

substitution and recycling, reduced consumption and device efficiency, urban planning, resource management, governance, and financing.

Dubash et al. (2013) propose and develop a methodology for operationalizing a co-benefits approach to climate policy formulation. They use the technique of multi-criteria analysis (MCA), which requires making choices between and examining trade-offs across multiple objectives of policy, such as growth, inclusion, and environment. MCA is the general term for a family of analytical techniques that are particularly relevant when assessing likely policy outcomes relative to multiple objectives, when values and consequent prioritization across those values may differ, and where it is important to assess both quantifiable monetary impacts and unquantifiable impacts. The authors argue that adopting an MCA-based co-benefits approach will likely bring gains to both domestic policymaking and India's international climate stance. Domestically, this approach would increase the coherence of policymaking early in the decision process. Internationally, a well-specified co-benefits approach will be a necessary first step to articulating India's policy approach based on the centrality of the principle of "common but differentiated responsibility and respective capabilities". In addition, the authors develop a framework for consideration of implementation issues (Tables 3.1, 3.2, 3.3 and 3.4).

References

Dietz S, Stern N (2014) Endogenous growth, convexity of damages and climate risk: how Nordhaus' framework supports deep cuts in carbon emissions. Centre for Climate Change Economics and Policy, Working Paper No. 180

Dubash N, Raghunandan D, Sant G, Sreenivas A (2013) Indian climate change policy. Econ Polit Weekly 48(22):47–62

Hope C (2006) The Marginal Impact of CO_2 from PAGE 2002: an Int grated assessment model incorporating the IPCC's five reasons for concern. Int Assess J 6(1):19–56

Indian Institute of Management Ahmedabad (IIMA), National Institute of Environmental Studies (NIES), Kyoto University (KU), Mizuho Information and Research Institute (MIZUHO) 2009 Low Carbon Society Vision 2050: India. Collaborative report prepared by Indian Institute of Management Ahmedabad, India, National Institute of Environmental Studies, Japan, Kyoto University, Japan, and Mizuho Information and Research Institute, Japan. (November 2009)

Intergovernmental Panel on Climate Change (2001) Synthesis Report. Cambridge University Press, Cambridge, UK

International Energy Agency (2007) World Energy Outlook. OECD/IEA, Paris, France

Mendelsohn R, Dinar A, Dalfelt A (2000) Climate change impacts on African agriculture. Preliminary analysis prepared for the World Bank, Washington, District of Columbia, p 25

MoEF (2009) India's GHG emissions profile: results of five climate modelling studies. Climate Modelling Forum, supported by Ministry of Environment and Forests, Government of India, New Delhi, India. http://www.moef.nic.in/downloads/home/GHG report.pdf

National Council of Applied Economic Research (2009) Climate Change Impact on the Indian Economy—A CGE Modelling Approach. NCAER, New Delhi, India

Nordhaus WD (1991) To slow or not to slow: the economics of the greenhouse effect. Econ J 101 (407):920–937

Nordhaus WD, Boyer JG (2000) Warming the World: economic models of Global Warming. MIT Press, Cambridge, MA

Planning Commission (2014) The Final Report of the Expert Group on Low Carbon Strategies for Inclusive Growth. Planning Commission, Government of India, New Delhi, India. Available online at http://planningcommission.nic.in/reports/genrep/rep_carbon2005.pdf

Shukla PR, Garg A, Dholakia HH (2015) Energy-emissions trends and policy landscape for India. Allied Publishers

Stern N (2008) The American Economic Review, vol 98, No. 2. In: Papers and proceedings of the one hundred twentieth annual meeting of the American Economic Association

Tol RSJ (2002) Estimates of the damage costs of climate change, Part 1: benchmark estimates. Environ Resource Econ 21:47–73

Chapter 4
Low Carbon Pathways

Himanshu Gupta

Abstract The climate debate for India encompasses issues other than just energy choices and energy efficiency. It is an integrated puzzle around lifestyle choices, aspirations of 1.2 billion people, and informed actions on water, air quality, and climate front. Further, in the wake of the COP-21 talks at Paris and the already announced US-China agreement on peaking of GHG emissions, India's approach to climate mitigation and adaptation is keenly watched in the climate policy space globally. This paper outlines one such approach using the India Energy Security Scenarios-2047 tool of the NITI Aayog and explores India's emissions trajectory till the year 2047 if the past trends continue and the trajectory if low carbon energy choices are made. Further, the study states the interventions in the demand and the supply sector which can make this transition feasible. However, such a transition will not happen without making investments in technologies, programs and infrastructure and hence a likely estimate of the quantum of investments is also estimated till the year 2047.

4.1 Introduction[1]

The NITI Aayog, (the erstwhile Planning Commission), rolled out an Integrated Energy Policy (IEP)[2] in 2006 that defines the concept of Energy Security for India as providing "lifeline energy to all its citizens irrespective of their ability to pay as well meet their demand for convenient energy for citizens to satisfy their various needs at competitive prices, at all times, considering shocks and disruptions that can be reasonably expected". It further talks about exploring options for achieving

[1]This paper is an updated version of the ICRIER Working paper 305 by the same author. This paper is available at http://icrier.org/pdf/Working_Paper_305.pdf.
[2]http://planningcommission.nic.in/reports/genrep/rep_intengy.pdf.

H. Gupta (✉)
Stanford University, Palo Alto, USA
e-mail: himanshugupta19@gmail.com

R. Kathuria et al. (eds.), *Low Carbon Pathways for Growth in India*, India Studies in Business and Economics, https://doi.org/10.1007/978-981-13-0905-2_4

India's Energy Independence beyond 2050. The concept of Energy Security might seem to be about exploiting domestic resources maximally and delivering energy at the cheapest price to the consumer, but it is often described with the associated implications for food, air quality, and water security: in short, sustainability. Because of global and local repercussions, a country cannot simply aim for high levels of growth based on fossil fuels alone. Local effects are a result of carbon-intensive fuels used in transport, biomass incineration, etc. Particulate matter emissions from power plants located close to cities have worsened their air quality: Delhi, already, is one of the lowest ranking cities in the world with respect to air quality, as per a recent World Health Organization[3] report.

Unabated emissions of CO_2 globally have given rise to fears that we might be on an irreversible path to a world where temperatures are poised to rise by 4° above those at pre-industrialization level.[4] The effects of such a pathway are both gradual and inflectionary. While climate science is yet to understand completely the tipping points of global warming, the signs from recent events are not encouraging. Cyclone Hudhud resulted in damages of more than 11 billion dollars[5] to the economies of Andhra Pradesh and Orissa, the latter being the most mineral-rich state in India. Gradual impacts have started to be felt already in the form of uneven monsoons, gradual rise in sea levels, and disturbing patterns of jet streams leading to polar vortices in the US. Adherence to the UNFCCC's 2° pledge will require substantial efforts, and contributions from all nations, OECD and non-OECD, will be crucial. These contributions will raise are issues relating to equity that will be debated in the relevant forums; it is imperative for India, a tropical country, to plan a growth trajectory, which is not only secure but sustainable, especially when 80% of our industrial production, 70% of our buildings stock, and more than 80% of our vehicle fleet is yet to be bought over the next three decades.[6] The choices we make today will determine our future over the next 3–4 decades.

It is in this context, that the NITI Aayog launched the India Energy Security Scenarios-2047[7] (IESS-2047); the aim was to explore the choices available for achieving energy security before 2047 (one hundred years of independence), and come out with a Max Energy Security pathway using the publicly available web interface. The present paper uses Version 2 of the tool to explore low carbon choices for India and goes on to calculate the cost implications of those low carbon choices. The tool has been recently put online for stakeholders' comments and reviews.

The paper is built upon assumptions around population growth taken from Scenario-B of the Population Foundation of India and, economic growth

---

[3]Ambient (outdoor) air pollution in cities database 2014, World Health Organization.

[4]http://www.rediff.com/business/slide-show/slide-show-1-special-most-polluting-countries-in-the-world-india-ranks-3/20130808.htm.

[5]http://www.dnaindia.com/india/report-cyclone-hudhud-in-andhra-pradesh-and-odisha-caused-us-11-billion-worth-of-losses-united-nations-2064515.

[6]http://niti.gov.in/content/india_energy.php.

[7]http://indiaenergy.gov.in/.

assumptions are taken from the IESS-2047, Version 2. The user activity demands consistent with those growth assumptions have been taken from the tool as well.

4.2 Assumptions

GDP: GDP is assumed to grow at a CAGR of 7.4% from the base year 2012–2047, growing at 6.8% from 2012–17, increasing to 8.1% between 2017–22, peaking at 8.4% in the between 2037–42, and coming down to 5.8% during the terminal years, 2042–47, of the study.

Structure of the economy: In line with the PM's vision of creating 30 million jobs in the manufacturing sector, the share of manufacturing in the GDP will have to rise: the share of manufacturing increases to 34% in the study in the year 2047, from the present levels of 16%.

Urbanization: In line with UN projections and the rising per capita incomes of Indians, urbanization will increase to 51% in 2047 from the present level of 31%.

Households per capita: Occupancy of households is a function of urbanization, and, in turn, income levels. Urbanization patterns for Indian states have been determined using census data. Tamil Nadu is the most urbanized state in India with 49% urbanization in 2011, and 4.1 people per household. The same has been regressed with urbanization data to arrive at a household occupancy of 3.9 for India in the year 2047. **The base year occupancy rate is 5.1 per household for India.**

Activity demand in the economy:

India's economy is assumed to grow at 7.4% CAGR from 2012–47. The economic activity so generated will lead to a demand for energy as well. The amount of economic activity generated over the 35 years is shown in Table 4.1.

The above level of economic activity will generate a large demand for energy four times that of base year levels as shown in the projections later. Fossil fuels cannot alone lead a secure and sustainable growth. Low carbon alternatives might seem expensive today but will bring benefits in the long term—even today, some low carbon technologies compete with conventional sources of energy. The price of electricity generated by wind turbines at some locations in India is comparable to that of electricity produced by gas turbines or imported coal-based electricity. If we add the cost of externality of burning coal and petroleum, this price can compete even with domestic coal-based electricity. An analysis by the IMF, quoted by the New Climate Economy mentions \$5–6/gigajoule as the externality cost for coal and \$0.4–\$0.6 per liter as the cost for diesel in India.[8]

In this context, future energy strategies for India in the long term cannot remain dependent on fossil fuels, which currently dominate the primary energy supply mix. For example, 46% of the primary energy supply required in 2011–12 was met by

[8]Brahmbhatt and Kathuria 2015. India: Pathways to Sustaining Rapid Development in a New Climate Economy: Conference Draft, http://newclimateeconomy.report/india/.

Table 4.1 Activity demand in the economy

Indicator	2102	2047
Per capita transport demand (pkm)	5970	18,132
Per capita steel use (kg)	66	372
Per capita building space—residential (m^2)	10.8	39
Per capita building space—commercial (m^2)	0.6	5.9

Source Author's calculations

coal. This will grow to almost 50% in the decade of 2040, going by the reference scenario, which is close to the present policy scenario. This is a cause for concern for two reasons. First, India does not have enough coal to meet this requirement; its coal production will peak during 2037–2042 at a level of 1170 million tons per annum going by the reference scenario of the IESS-2047. Second, coal is the biggest contributor to emissions, as it has the highest emission factor and maximum usage among the fossil fuels. India cannot afford to have such a high coal footprint in its electricity-producing, industrial processes, and other emery-consuming activities.

Our strategy for a sustainable and energy-secure future should focus on phasing out fossil fuels, by electrification and, in turn, supply that electricity from clean sources of energy, which are developed domestically. However, clean sources of energy such as biofuels, solar, wind, and even hydrogen are currently costlier and require risky investment decisions to be taken now, in order for them to scale in the medium term and compete with conventional sources of energy in the long term. Another approach towards a sustainable and secure energy future could be through the introduction of energy efficiency measures in all energy demand sectors such as agriculture, industry, and transport, and by cutting down energy consumption significantly. This strategy is also capital intensive but yields results in the short-to-medium time frame. Energy efficiency interventions mitigate the risk of low carbon technology investments, many of which are not yet commercialized.

This paper aims to look at the reference scenario of the NITI Aayog using the publicly available IESS-2047 tool. With the help of experts and external consultation, it will arrive at the baseline low carbon pathway and the costs of achieving it. The cost of not achieving the pathway has been analyzed separately in this report.

4.3 Reference Scenario

This pathway envisages a growth rate of 7.4%. The government is expected to implement policies dealing with demand, while the deployment of technologies relating to supply continues to be in line with past trends. This pathway envisages a growth rate of 7.4%, with the population rising to 1.7 billion in the year 2047.

4.3.1 Demand

India's energy demand rises by a factor of 4 from the base level of 420 Mtoe in 2012, to 1725 Mtoe (Fig. 4.1) in the year 2047. The per capita primary energy demand grows less than three times from 346 Kgoe (2012) to 1013 Kgoe (2047). Buildings and industry demands rise by a factor of 10 and 4.4 between the base year and 2047, respectively. The economy moves towards electricity as a fuel with electricity accounting for 19% of the primary energy supply in the year 2047, driven by moderate renewables deployment in the power sector to a level of 2% (Fig. 4.1). This scenario would also entail ramping up of liquid biofuel production in the country, to 11.6 MT (Fig. 4.2). Within the transport sector, in a reversal of trends, 15% people travel by rail rather than road (2047), as compared to 14% in 2012. Most of the demand in the buildings sector is fuelled by the rapid growth in housing both, in urban areas by rise in income levels leading to a more rapid

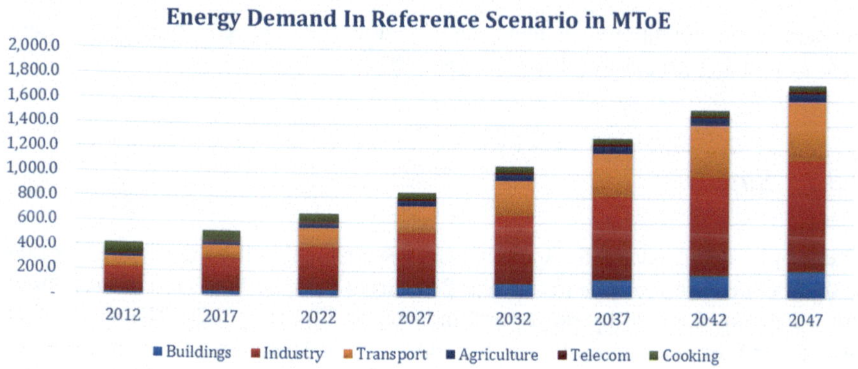

Fig. 4.1 Composition of the demand sector in the reference scenario

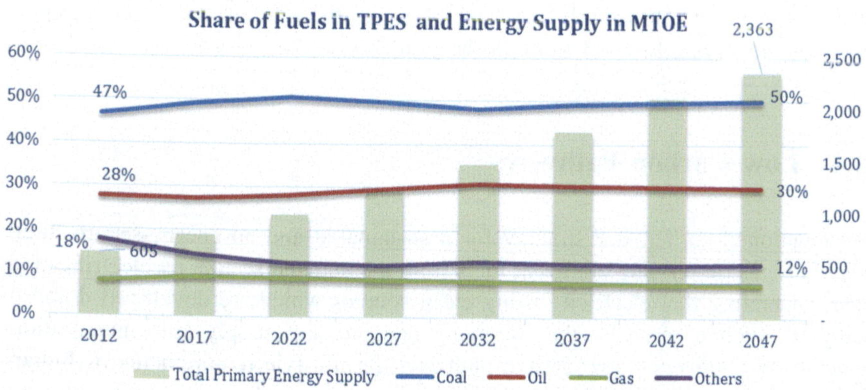

Fig. 4.2 Composition of the supply sector in the reference scenario

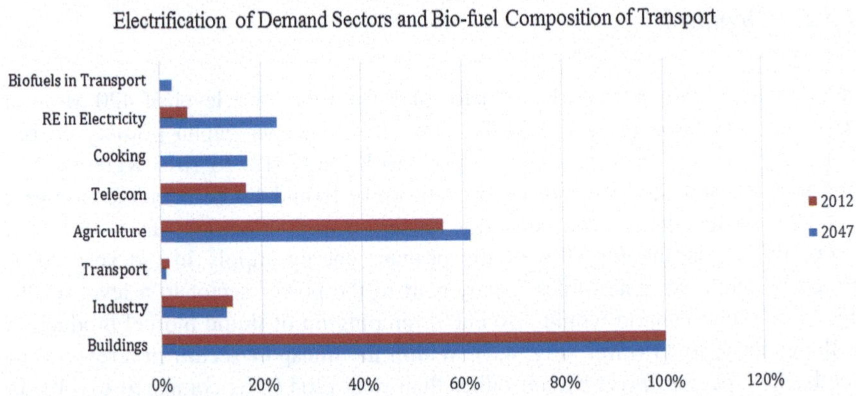

Fig. 4.3 Electrification* of the demand sectors in the reference scenario. *The definition of buildings in this case does not include the cooking sector and includes building envelope, insulation and appliances

increase in air conditioner demand, and in rural areas, by government programs such as housing access for all by the year 2022.

4.3.2 Supply

The supply side equation for India's energy sector will still be dominated by coal with its share rising to 50% in the year 2047 from the base level of 47%, driven by huge demands for coal by power and industry sectors (Fig. 4.2). The share of oil rises to 30% driven mainly by the transport sector. The share of electricity in the primary energy supply increases to 19% in 2047 from the present level of 15% (Fig. 4.3). Renewable penetration in the grid increases to 23% in the year 2047, requiring smart grid reforms and considerable investments in grid and balancing capacity to be able to absorb this. Consequently, the total primary energy supply increases from 605 to 2363 Mtoe in the year 2047.

4.4 Low Carbon Pathways

As mentioned earlier, our strategy for a sustainable and an energy-secure future should focus on phasing out fossil fuels from the demand sectors by electrification, and, generating that electricity using clean sources which are developed domestically. In such a scenario, the following demand and supply side interventions emerge. In choosing a low carbon pathway, the lifestyle requirements of Indians have not been moderated, therefore ownership patterns of Cars in Transport Sector,

Buildings, Material Consumption, both remain the same. However, it is assumed that changes in the behavior of consumers, penetration of clean technologies such as electric vehicles in the transport sector, and in the electricity sector will steer us towards the low carbon pathway by 2047.

In this pathway, every sector contributes to both energy security and environment sustainability. Due to heroic efforts, demand is reduced to the minimum possible in 2047. On the other hand, the supply has been calibrated so as to minimize emissions. GHG emissions rise moderately from 1.7 ton/capita to 3.6 tons/capita by 2047 (population assumed at 1.7 billion), in spite of a tripling in per capita energy supply (Fig. 4.4). Heroic efforts will also entail electrifying demand sectors to the maximum technical, economic, and behavioral limit and supplying the electricity required using renewable sources. Also, coal- and gas-based thermal capacity additions are assumed to peak by 2032. Renewables are ramped up in order to meet the targets of 100 GW for solar and 65 GW for wind as was also announced by the Prime Minister.[9] It is assumed that biofuels, second generation, advanced and algal based, will be ramped up by government support, and breakthroughs in commercialization to an extent of 110 MT of oil equivalent.

4.5 Demand

India's energy demand rises by a factor of 3 from the base level of 420 Mtoe in 2012 to 1347 Mtoe (Fig. 4.5) in the year 2047. The per capita demand of energy rises by merely a factor of 2 from the present level of 346–790 Mtoe in 2047. Demand from transport and industry increases by a factor of 6.9 and 3.4 respectively, in 2047 as compared to 4.4 in the reference scenario in both the cases. The share of electricity in the primary energy supply increases to 25% in the year 2047, due to electrification of the demand sectors—industry, 22%, and cooking 26%, among others (Fig. 4.6). In the transport sector, in 2047, 80% of the two-wheelers and 38% of the cars run on electric engine technology.

On the supply side, the share of coal will come down to 34% in 2047, becoming similar to the share of other fuels (renewables, clean energy and bioenergy) in the system (Fig. 4.6). The share of electricity in the primary energy supply will rise to 22% in 2047, and the share of renewables in the electricity grid would be 56% in the year 2047 (Fig. 4.7), the highest among all the scenarios. Solar and wind power completely satisfy the summer and monsoon afternoon peak (albeit some curtailment might become necessary). Grid balancing support in these seasons is mainly the capacity support (ramping, reserves, and other ancillary services). There has been a deal on an interconnecting grid between SAARC nations off-late. This, if it comes online, might offer additional leeway for excess generation through renewables and reduce the need for curtailment by an equivalent amount. Biofuels

[9]http://pib.nic.in/newsite/PrintRelease.aspx?relid=122567.

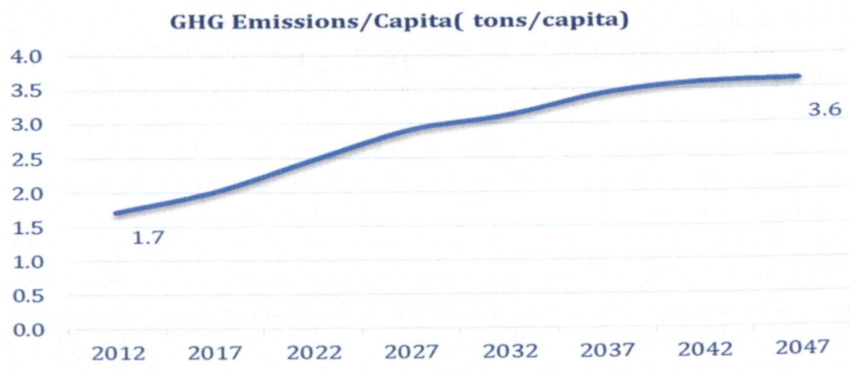

Fig. 4.4 Emissions per capita in the low carbon scenario

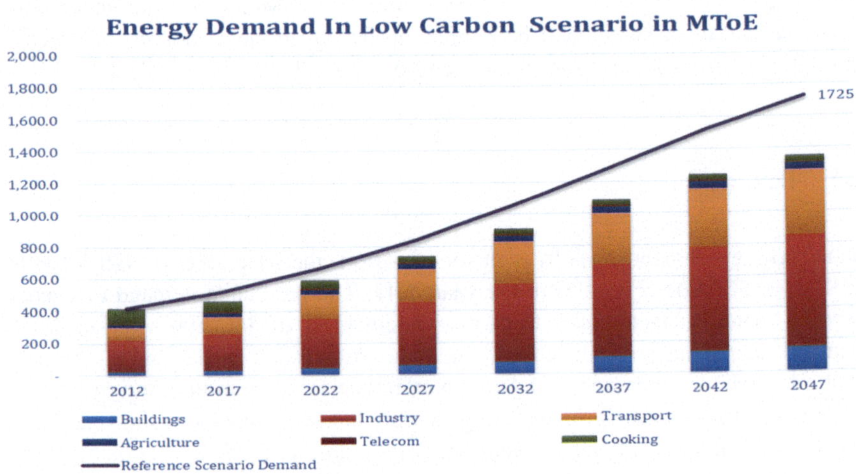

Fig. 4.5 Composition of the demand sector in the low carbon scenario

will meet 30% of the demand for liquid transport fuels (Fig. 4.7), which implies massive investments in the R&D of second generation and advanced biofuels. As a result, emissions would be lowest across all scenarios, with emissions per capita peaking at 3.6 tons. A analysis of cumulative emissions in both the pathways—reference and low carbon is presented in Fig. 4.8.

A first look at the graph clearly shows that emissions in the minimum emissions or low carbon pathway will be a notch higher than the anticipated target of 2 tons/capita as advocated by the UNFCCC for preventing a temperature rise beyond two degrees. That said, there are equity issues associated with the fair allocation of carbon budgets based on historical cumulative emissions which are likely to be resolved in appropriate forums. The emissions analysis for the 2012–2047 period of the study does not take into account historic emissions. In any case, this transition

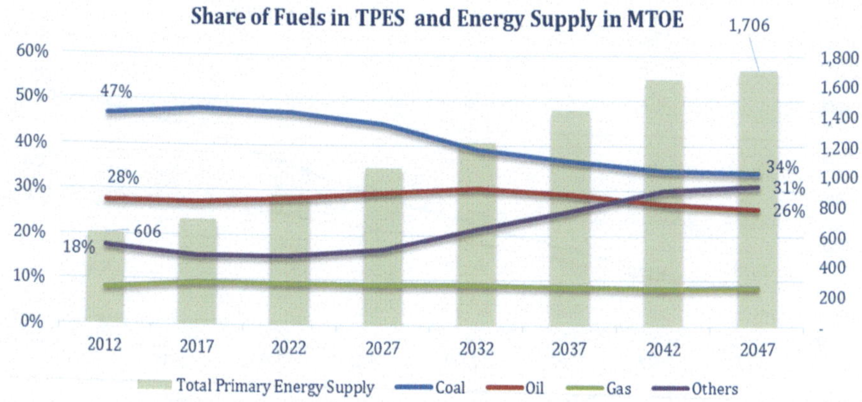

Fig. 4.6 Composition of the supply sector in the low carbon scenario

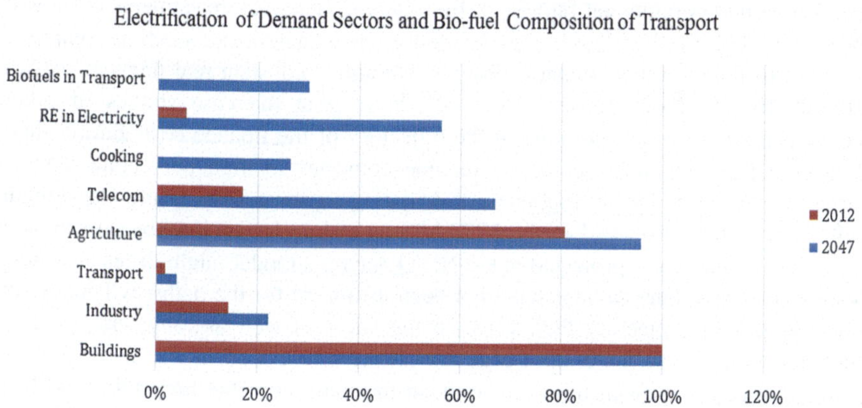

Fig. 4.7 Electrification of the demand sectors and biofuel composition of the transport in the low carbon scenario

would come at a huge cost to the economy. This paper tries to explore the feasibility of various low carbon choices and tries to estimate the cost of transitioning for India to a low carbon pathway. For the purposes of this paper, we have taken into account the capital investments required in various demand sectors to realize energy efficiency. The paper refrains from exploring the mitigation potential of CCS as we do not believe that it will make a dent in industry scenario as projected by studies such as the IEA New Policy Scenario.

Before interpreting the table, it is worthwhile to note that the investment costs have been annualized to calculate the cost of energy efficiency interventions in the demand sector. Each sector's costs include capital and finance cost. Capital costs refer to the costs of electric vehicles, electric locomotives, and insulation in

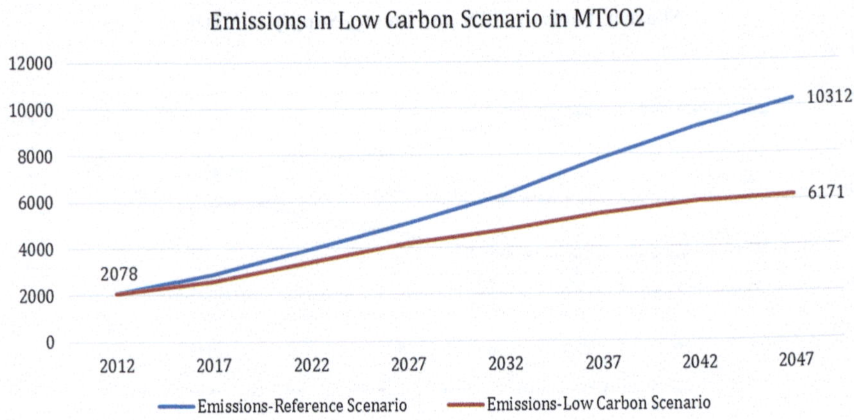

Fig. 4.8 Emissions trajectory in ref and low carbon scenario

residential and commercial buildings, for example. The real investments costs will be even higher. Further, these costs are relative costs/incremental costs as compared to those of the reference scenario. We have assumed constant real finance costs of around 4% for India between 2012–47, looking at the case studies of other developed countries and looking at the trajectory of the finance costs during their transition from medium per capita income countries to high per capita income countries. Some of the assumptions in calculating capital and operating costs might differ from those projected in the IESS-V2 put online for review, because expert opinion held that costs projected in the NITO Aayog's model might differ in sectors such as biofuels. Fuel savings have not been estimated for the pathway because of uncertainties in projecting their price in the short as well as long term. Recent movements in the prices of spot coal and oil are a reflection of the uncertainty in estimating prices even in the short term; in the long term, the uncertainty will be even higher, as European coal demand is flat and coal demand from China is likely to have peaked last year.[10] Moreover, these fossil fuel savings will be realized over a period beyond 2047. However, looking at India's development narratives, investments are required to be made now.

Table 4.2 indicates that the low carbon pathway will cost $ 2030 billion for India during the period 2012–47, which is 0.7% of the GDP in the same period. These costs do not take into account the savings from reduced consumption of fossil fuels and benefits resulting from improved air quality and water savings. Close inspection reveals that almost 18% of the pathway costs will be incurred on biofuels, second generation and advanced. These technologies are still in the laboratory and significant investments will be required in R&D and their commercialization. 31% of the costs will be incurred on electricity, and the pathway talks about 55% penetration of renewable electricity into the grid. Costs in the transport sector are

[10]http://thinkprogress.org/climate/2015/05/27/3662681/chinas-coal-use-peaked/.

Table 2 Investments cost of low carbon interventions in billions of dollars (2012–47) over reference scenario

Sector	Investment costs (USD Billion)	% Share in the overall investment costs
Buildings	308	15%
Transport	302	15%
Industry	157	8%
Bioenergy and others	366	18%
Electricity	631	31%
Finance	266	13%
Total	**2030**	**0.7% of GDP(2012–47)**

Source Author's calculations

moderate compared to those in the other sectors, and are primarily incremental investments that will give momentum to Electric Vehicles and Fuel Cell Vehicles in the country to the desired levels of penetration. Cumulatively, these investments will amount to 0.7% of the cumulative GDP between 2012–47 at 2012 prices.

4.6 Conclusion

1. The climate debate for India encompasses issues other than just energy choices and energy efficiency. It is an integrated puzzle around lifestyle choices, aspirations of 1.2 billion people, and informed actions on the water, air quality, and climate fronts. We have not dealt with air quality and water issues in this study and the conclusions are limited to impacts on emissions and energy demand/supply. Further, the paper evaluates energy-related emissions as outcomes. Land use and livestock-related emissions were not modeled in this study.
2. More than 90% of steel till 2047 is yet to be produced and more than 80% of our residential building stock is yet to be constructed giving us a window of opportunity to make informed choices now. These choices will determine our future for the next 50 years taking into consideration the asset life of the infrastructure that needs to be created in order to meet these lifestyle choices.
3. In case of the reference scenario, the emissions rise from the base level of 2 GT to the 2047 level of 10.3 GT CO_2 equivalent, the current level of China's annual emissions. With heroic efforts around energy efficiency interventions in the demand sector and a shift to low carbon choices in the supply sector, these emissions could be limited to 6.1 GT in 2047. In evaluating these choices, we have kept the lifestyle ambitions of Indians constant across both the scenarios. As an example, ownership level of cars is not only an outcome of income levels but also the society structure and culture issues, which cannot be modeled.

Any interventions in those areas and the likely response of Indians to those policy interventions have been kept outside the scope of this study.

4. The level of 6 GT is still higher than the envisaged levels of 2 GT, but any further effort to mitigate emissions to below 6 GT requires changes in lifestyle, changes in the way we consume materials and design buildings, which other studies have tried to model. Many equity issues are associated with the envisaged level of 2 GT which needs to be resolved in the appropriate forums.
5. The investment cost of bringing down emissions to 6 GT is likely to be around 2.03 trillion dollars, 0.7% of GDP from the year 2012–2047. Within that, 31% of investments will need to be made in electricity sector, primarily in renewables, and 18% in the bioenergy sector. The investments required in the demand sectors—transport, buildings, and industry—are likely to be less than those in the supply sector: 15, 15, and 8%, respectively. It may be noted that these costs should be used as indicative costs and their determination is highly uncertain.

Chapter 5
Cost of Inaction on Mitigating Climate Change: A Preliminary Analysis

Vaibhav Chaturvedi

Abstract India is one of the most vulnerable countries to climate change impacts. Climate change impacts are many and varied, and this paper provides only first-order approximations. In the agriculture sector, the paper has looked only at output losses of three major food crops—rice, maize and wheat. While the health impacts of climate change include mortality at old age due to heat waves, this paper has focused on deaths related to three important diseases—diarrhea, malaria, and dengue. Impacts on energy infrastructure will be many, and the analysis has focused on increased requirement of power generation for meeting peak hour demand of electricity.

5.1 Introduction

The Fifth Assessment Report of the IPCC (2014) has reiterated that climate change is real and its impact is being felt across the world. Mitigation action is immediately required to limit the atmospheric concentration of greenhouse gases. Mitigation implies shifting away from the current energy system to fundamentally different decarbonized energy systems, and this shift entails cost. Mitigation costs are holding most governments away from investing in emission mitigation efforts at the scale and speed required to combat climate change. In absence of this investment, climate change is bound to happen, and the cost of climate change impacts will be increasingly borne by the world.

The influential study led by Dr. Nicholas Stern and also known as the Stern review (Stern 2006) was instrumental in highlighting the cost of climate change impacts. Impacts are varied in terms of their nature as well as intensity. Increased temperatures are expected to reduce agriculture productivity, increase incidences of vector-borne diseases, impact the hydrological cycle, biodiversity, and ecosystems,

V. Chaturvedi (✉)
Council on Energy, Environment and Water (CEEW), New Delhi, India
e-mail: vaibhav.chaturvedi@ceew.in

R. Kathuria et al. (eds.), *Low Carbon Pathways for Growth in India*, India Studies in Business and Economics, https://doi.org/10.1007/978-981-13-0905-2_5

and also lead to an increase in the frequency and intensity of extreme events such as cyclones.

Climate change mitigation is a global challenge, but its impact will vary across regions and temperature zones. Small island states will be hit the hardest with sea level rise. Bigger countries, such as India, are expected to be more vulnerable owing to its large agricultural sector, vast population, rich biodiversity, long coastline, and high poverty levels. Also, India has pushed for inclusion of adaptation as a part of Intended Nationally Determined Contributions (INDC). For understanding adaptation requirements, we need to understand and value climate change impacts first. This short assessment tries to estimate the cost of global climate change mitigation inaction on India. We aim at estimating first order costs for loss in agricultural productivity, health impacts, and the impact of increasing temperatures on increased power generation requirement.

5.2 Methodology

Our approach estimates the cost of key impacts for the years 2050 and 2100. The rationale for choosing the main three impacts for cost assessment is the following: (i) India's agriculture sector is the source of livelihood for more than 65% of its population, and agricultural productivity is considered to be low compared to global agricultural productivity. Moreover, the Indian government's aim is to always be self-sufficient in terms of food production. Given these realities, any decline in agricultural production is bound to be costly for the nation and this cost needs to be assessed; (ii) With a huge population exposed to health impacts due to low resilience and income, any increase in chances of negative health impacts from increased incidence of vector-borne diseases will pose additional challenges. Health is a social concern and health provision will be further challenged due to impacts from climate change; (iii) Extreme temperatures will, in all likelihood, increase for any given year, which will increase cooling requirements. This is an energy sector impact as additional power plants will need to be installed, and it is important to know what this cost will be for India.

The choice of the above mentioned impacts does not mean that other impacts are negligible but reflects a high degree of uncertainty in data related to either biodiversity loss or increase in intensity and frequency of extreme events. Within these sectors, we limit ourselves to some key categories, so as to be able to present indicative numbers and approximations upon which further discussions and studies can be based.

For understanding the cost of climate impacts on agriculture and additional power plant generation, outputs from the Global Change Assessment Model (GCAM) are used (Fig. 5.1). GCAM is a global integrated assessment model with separate agriculture and land use system components (Clarke et al. 2008; Calvin et al. 2009; Wise et al. 2009; Shukla and Chaturvedi 2012; Chaturvedi et al. 2013a, 2014a). Information on total production of rice, wheat, and maize in 2050 is based

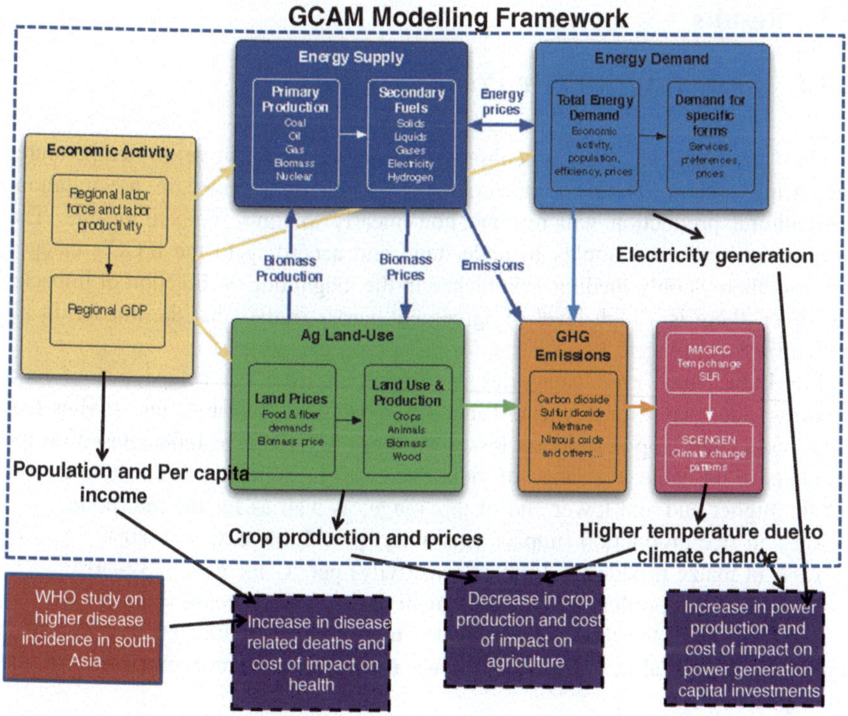

Fig. 5.1 Methodological framework for understanding cost of inaction across key sectors

on the business-as-usual (BAU) model run. Literature is reviewed to understand the rate of decline in crop productivity due to an increase in temperature in 2050 and 2100. On the basis of this information, total crop losses have been identified and valued, based on prices in the respective years. Details on GCAM's agriculture and land use module can be found in Wise et al. (2009) and Chaturvedi et al. (2013b).

The GCAM also models cooling and heating demand based on cooling/heating degree days and a host of other factors. This modeling analysis will give information on whether increased energy demand for cooling will imply a significant cost for India or not for additional power generation infrastructure. Details of GCAM's building sector module can be found in Chaturvedi et al. (2014b).

Finally, health impacts are determined by linking increasing temperatures to increased incidence of diseases and what it means in terms of additional health costs based on the literature. The WHO has already done a detailed quantified assessment for the world and various regions (WHO 2014a). This analysis borrows results for South Asia from the WHO study and derives India-specific health impact numbers based on the South Asian results.

5.3 Results

5.3.1 Cost of Agriculture Production Loss

Studies have shown that agricultural production is sensitive to temperature, increasing carbon dioxide concentrations as well as changes in precipitation. Agricultural production will respond non-linearly to future climate change. The impact, however, is complex to understand and according to the IPCC's categorization, there is only medium confidence in the magnitude or direction of impacts. However, there is a high level of agreement across studies that the impact, in all probability, is going to be negative for most crop categories.

For India, three crop categories are important from the perspective of food security: rice, maize, and wheat. Table 5.1 shows results from a few studies that have researched crop production losses for these key crops in India. Based on the numbers in the table, the range of yield decreases is estimated. We estimate costs for the higher end and lower end of this range, as well as for the mid-point.

Loss in rice production (impact sensitivity) per °C increase = 4–20%

Loss in maize production (impact sensitivity) per °C increase = 32–50%

Loss in wheat production (impact sensitivity) per °C increase = 5–20%

It should be noted that these estimates include not just impacts due to higher temperatures but also impacts of higher carbon dioxide concentrations in the atmosphere.

We use the following formula to calculate the impacts of climate change on the three major Indian food crops:

$$CoI_{Ag,Y}=\left\{Pdt_{BAU,Y}*\left(1-[1-ImS]_{Y}^{Temp}\right)\right\}*P_{BAU,Y}$$

where

CoI	is the cost of Inaction in Million US$
Pdt	is production in Million Tones
Temp.	is the temperature increase relative to BAU in °C
ImS	is impact sensitivity of crop production to increase in temperature in %/°C
P	is price in US$/ton

The subscript Ag denotes agriculture, and subscript, *Y*, denotes the year under analysis; BAU stands for business as usual.

Crop production, temperature, and crop prices are outputs of the GCAM. Impact sensitivity is a crop-specific constant derived from the literature as highlighted above. Temperature increase has been taken relative to 2005, which is the model base year. The function in curly brackets represents physical loss of production, which when multiplied by the price gives the cost of loss in agricultural output. Table 5.2 shows the loss in physical production as well as in terms of economic losses.

Table 5.1 Decline in production of rice, maize, and wheat in India due to climate change

Year	Crop	Loss in production with approx. 1° rise in temperature	Region	References
NA	Rice	−20%	India	Senapati et al. (2013)
2030	Irrigated rice	4% loss in production in majority of districts	Western Ghats	Kumar et al. (2011)
2030	Irrigated rice	10% loss in production in majority of districts	Coastal Districts	Kumar et al. (2011)
2030	Irrigated rice	5% increase in production in majority of districts	North-East India	Kumar et al. (2011)
2030	Rainfed rice	10% loss in production in majority of districts	Western Ghats	Kumar et al. (2011)
2030	Rainfed rice	0% (midpoint value)	Coastal Districts	Kumar et al. (2011)
2030	Rainfed rice	10% loss in production in majority of districts	North-East India	Kumar et al. (2011)
2030	Maize	50% loss in production in majority of districts	Western Ghats	Kumar et al. (2011)
2030	Irrigated maize	32% loss in production (midpoint value across sub regions)	Coastal Districts	Kumar et al. (2011)
2030	Rainfed maize	35% loss in production in majority of districts	Coastal Districts	Kumar et al. (2011)
2030	Irrigated maize	40% loss in production in majority of districts	North-East India	Kumar et al. (2011)
2030	Wheat	20% loss in production in majority of districts	North-East India	Kumar et al. (2011)
2020-30	Wheat	4–5 million tons with 1° rise, relative to base year conditions. (In 2008 publication year, actual production was 78 mn ton. Implies 6% loss approx.)	India	Aggarwal (2008)
2004	Wheat	4 mn ton with 1° rise (actual production in 2004 was 72 mn ton. Implies 5.3% loss approx.)	India	Samra and Singh (2004)

Source: Author's calculations

Table 5.2 Cost of impacts on agriculture sector food crops

	Pdt (Mn/Ton)	Increase relative to 2005	Impact sensitivity (% per °C)			Global CropPrice	Loss in output (million tonnes)			Percentage loss in output (Relative to BAU)			Cost of impacts-million US$ (2010 prices)		
	Millon					2010									
	Tonnes	°C	Low	Medium	High	US$/kg	Low	Medium	High	Low	Medium	High	Low	Medium	High
2005	–	–	–	–	–	–	–	–	–	–	–	–	–	–	–
Rice	136	0	4%	12%	20%	–	–	–	–	–	–	–	–	–	–
Maize	15	0	32%	40%	50%	–	–	–	–	–	–	–	–	–	–
Wheat	70	0	5%	12%	20%	–	–	–	–	–	–	–	–	–	–
2050	–	–	–	–	–	–	–	–	–	–	–	–	–	–	–
Rice	199	1.46	4%	12%	20%	2.26	11.51	33.88	55.33	6%	17%	28%	25,997	76,499	124,931
Maize	24	1.46	32%	40%	50%	1.45	10.33	12.62	15.28	43%	53%	64%	14,988	18,299	22,159
Wheat	115	1.46	5%	12%	20%	1.92	8.30	19.58	31.97	7%	17%	28%	15,900	37,518	61,270
–	–	–	–	–	–	–	–	–	–	–	–	–	56,886	132,317	208,360
2100	–	–	–	–	–	–	–	–	–	–	–	–	–	–	–
Rice	199	3.26	4%	12%	20%	2.18	24.80	67.82	102.86	12%	34%	52%	53,994	147,679	223,967
Maize	27	3.26	32%	40%	50%	1.36	19.32	21.89	24.18	72%	81%	90%	26,278	29,777	32,889
Wheat	121	3.26	5%	12%	20%	1.76	18.63	41.24	62.54	15%	34%	52%	32,707	72,388	109,783
													112,97	249,844	366,639

Source Author's calculations

The literature shows that maize is likely to be impacted the most because of temperature increase, followed by rice and wheat. A 3.25°C increase in average temperature by the end of the century, relative to 2005, can result in a decline of 72–90% in the output of maize; 12–52% in the output of rice and 15–52% in the output of wheat. The total economic loss is US$57-208 bn in 2050 and US $113-367 bn in 2100. In terms of GDP share, the economic losses from these three crops amount to 0.28–1.02% in 2050, and 0.26–0.84% in 2100.

5.3.2 *Cost of Health Impacts*

Diarrhoea-related child mortality

A recent report by the World Health Organization (WHO 2014a) suggests that increasing temperature will increase the rate of spread of diarrhoea-related deaths. The study uses the following function to estimate climate attributable diarrhoeal deaths:

$$n_{c,y,i,j} = N_{c,y}\frac{\exp\left[\beta i\left(\Delta \mathrm{T}_{c,y,j}\right)\right] - 1}{\exp\left[\beta i\left(\Delta T_{c,y,j}\right)\right]}$$

where

- N is the number of climate attributable diarrhoeal deaths,
- N is the number of diarrhoeal deaths without any climate change, for reference,
- T is the change in temperature with climate change relative to fixed climate,
- β denotes the sensitivity of diarrhoeal death to temperature increase, and is calculated as $\beta = \log(1 + \alpha)$, where α is the linear increase in diarrhoeal deaths per degree rise of temperature,

Subscript '*c*' denotes a grid cell, '*y*' denotes time slice, '*j*' represents three different scenarios of temperature anomaly, and '*i*' denotes low, medium, or high level of diarrhoea-related deaths.

Malaria-related mortality

Malaria is a disease that has shown a drastic decline with time as incomes across countries have risen. However, in low-income countries of the world this is still the case. Interestingly, in India, malaria-related cases were reported to be around 2 mn in the 1990s, though the WHO estimated this figure to be 15 mn (Kumar et al. 2007). The WHO (2014a) has also estimated that with increasing incomes and no climate impacts, malaria will be eliminated from all the regions of the world except Africa, by 2050.

The WHO (2014a) uses a regression equation to estimate the impact of increasing temperatures, changing precipitation and increasing income on the risk of population exposed to malaria.

$$\text{logit}(\text{Malaria}_1) = \beta_0 + \beta_1 T\_\text{min}_1 + \beta_2 PR\_\text{max}_1 + \beta_3 \sqrt{(\text{GDP}_{pc_1})}$$

where

*T_*	min is the mean temperature of the coldest month,
*PR_*	max is the mean precipitation of the wettest month,
GDP	is the GDP per capita,
'*i*'	denotes spatial grid location

Dengue-related mortality

Dengue fever is a vector-borne disease. Climate affects dengue at a high rate in tropical regions as the transmission capacity increases. It is a disease that has shown a drastic decline with time as incomes across countries have risen. As in the case of malaria, there are many factors that influence the spread of dengue and hence the impact of climate change is uncertain at best.

WHO (2014a) uses a regression equation to estimate the impact of increasing temperatures, changing precipitation, and increasing income on the risk of population exposed to dengue.

$$\text{logit}(\text{Dengue}_i) = \beta_0 + F(\text{Temperature}_i, \text{precipitation}_i) + \beta_1 \text{GDP}_{pc_i}$$

where

Temperature	is the annual mean temperature
Precipitation	is the annual mean precipitation,
F	is a Spline function,
GDP	is the GDP per capita,
'*i*'	denotes the spatial grid location

Cost of climate change induced deaths

The WHO (2014a) estimates are based on sophisticated modeling at the grid level across various regions of the world. However, results are presented only for South Asia. This study assumes that for 2050, India will have the same share of climate-induced deaths as south Asia for all the diseases being analyzed. For 2100, we use assumptions based on the 2030 and 2050 share of deaths as modeled for South Asia.

For getting from number of deaths to cost of deaths, we have to put a value on the life of a person, which is a daunting task. Though we believe that one value cannot be put to any life, we make some assumptions for the purpose of our calculations. We assume that any life lost leads to a loss in GDP, equal to income forgone for 50 years of work life. For putting a value to life in 2050, we add per capita income for India from 2005 to 2055, which signifies total income for a person across his or her work life. In other words this is the income forgone when a life is lost. For value of one life in 2100, we use the same approach and per capita income is added from 2050 to 2100. Table 5.3 describes our assumption, calculations, and the final result.

Table 5.3 Cost of impacts on human health due to higher disease burden

	2030	2050	2100	Source
South Asia				
Population (Million)	2749.43	3188.78	–	GCAM
Diarrheal Deaths	14,870	7717	–	WHO (2014a)
Malaria Deaths	1875	9343	–	WHO (2014a)
Dengue Deaths	39	209	–	WHO (2014a)
Diarrheal Deaths as a %	0.00054	0.00024	0.00005	2100 value is our assumption based on 2030 and 2050 values
Malaria Deaths as a %	0.00007	0.00029%	0.00126	
Dengue Deaths as a %	0.000001	0.00001	0.00003	
India				
Population (Million)	–	1736	1552	GCAM assumption
Diarrheal Deaths	–	4201	776	Based on WHO (2014a)
Malaria Deaths	–	5086	19,537	Percentages calculated for South Asia have been multiplied by Indian population
Dengue Deaths	–	114	470	
Assumed value of life (US$, 2010 prices)	–	–	–	Based on per capita GDP in GCAM
	–	265,000	10,62,000	
Value of lives lost due to climate	–	–	–	
change induced effects	–	–	–	
Diarrheal Deaths (Million US$, 2010 prices)	–	1113	824	
Malaria Deaths (Million US$, 2010 prices)	–	1348	20,748	
Dengue Deaths (Million US$, 2010 prices)	–	30	499	

Source Author's calculations

As is evident, the results include the positive impact of rising incomes in India across the century. Diarrhoeal risk should be eliminated by 2050 if there are no climate change impacts. However, climate change does lead to increase in deaths compared to the no climate change scenario. Most importantly, deaths related to malaria are bound to increase significantly and the resulting loss of economic output is US$20.7 bn. in 2100 for malaria alone.

5.4 Increased Investment in Electricity Generation Infrastructure

Climate change induced temperature increase is bound to increase space cooling demand in both residential and commercial sector. The GCAM uses a detailed approach including technical and economic factors for modeling space cooling demand (Chaturvedi et al. 2014b). The following functional form is used for modeling cooling service demand:

$$d_C = k_C(\mathrm{CDD}_{\eta r} + \lambda_C \mathrm{IG})\left[1 - \exp\left(-\frac{\ln 2}{\mu_C}\frac{i}{P_C}\right)\right]$$

where

d_c is the demand for cooling service per unit floor space in EJ-output/m^2
CDD is cooling degree days in hr °C which changes over time,
η is the thermal conductance or building U-value in GJ/m^2 h^{-1}°C^{-1},
r is the building floor-to-surface area ratio representing the size of building shell exposed to outdoor temperature,
IG is the amount of building internal gains in GJ/m^2, and,
λC is the internal gain scalar accounting for the potential mismatch of the time when space conditioning is required and the time when the internal gains are produced,
i is per capita income,
P_c is the price of cooling service, which is endogenously determined,

μc represents the parameter determining speed with which service demand increases in response to change in income and prices towards the satiation level.

The term "CDD" is what changes between a fixed climate and a changing climate. Fixed climate represents CDD for 2005, while changing climate corresponds to the higher temperature increase, close to 4 °C observed by the century end. The model does not model peak and base load demand separately and treats all the technologies equally, which can be regarded as a limitation. Hence, in this model, increased demand for electricity production is distributed between different technologies such as coal, gas, nuclear, solar, etc. on the basis of relative cost dynamics. However, it is assumed here that all this increase will be for meeting peak energy demand and hence a gas-based power plant is most suitable for meeting peak power demand. On the basis of the GCAM's output as well as capital cost assumptions based on the Annual Energy Outlook (AEO 2013), we calculate the increase in investments required related to power plants (Table 5.4).

Temperature-induced higher peak load and cooling energy demand will lead to additional installed capacity of 36 GW in 2050 and 136 GW in 100. Total generation capacity needs to increase 9–10-fold by 2100 for power consumption to equal that of average developed country levels, or above 2000 GW of installed capacity in the distant future across all technologies (nuclear, solar, coal, etc.). If the

Table 5.4 Cost of impacts on power sector for meeting higher peak energy demand

	2050		2100		Source
	Fixed climate	Changing climate	Fixed climate	Changing climate	
Electricity production (EJ)	27.41	27.87	34.78	36.50	GCAM
Increase in production (EJ)	–	0.46	–	1.72	GCAM
Conversion: KWh/GJ	–	277.78	–	277.78	–
Increase in production (GWh)	–	128846.80	–	476942.82	–
Gas power plant capacity factor	–	0.40	–	0.40	Assumption
Hours in a year	–	8760.00	–	8760.00	–
Increase in production (GW)	–	36.77	–	136.11	–
Capital cost of gas power plant	–	–	–		–
(US$/KW of installed capacity, 2010 prices)	–	-905.00	–	905.00	AEO (2013)
Total additional investment (Million US$, 2010 prices)	–	33,278	–	123,183	–
	–	35	–	–	–

Source Author's calculations

additional power demand is met by a technology with higher average capacity factor, say coal, then the additional installed capacity will be much lower. However, it makes most sense to install gas-based power production for meeting additional peak load demands as this technology gives low-cost flexibility to meet hourly power generation requirements.

5.5 Conclusions and Limitations

The Stern Review (Stern 2006) highlights the fact that the total cost of climate change under the BAU scenario is estimated to be at least 5% of the value of global per capita consumption over the next two centuries. Indeed, India is one of the countries most vulnerable to climate change impacts. Climate change impacts are many and varied, and the present analysis only offers limited initial insights. The study aims at only first-order approximations, and the motivation for this analysis is to start a wider discussion for a more robust assessment of climate impacts and their valuation across sectors in India, all within the same analytical framework. Hence in the agriculture sector, the study has looked only at output losses of three major food crops: rice, maize, and wheat. However, climate change will impact all categories of crops ranging from oilseeds to fruits and vegetables. Health impacts of climate change include mortality at old age due to heat waves, higher incidence of

Table 5.5 Summary of cost of inaction across sectors

	Cost of inaction in absolute terms			As percentage of GDP	
	2050	2100		2050 (%)	2100 (%)
GDP	20,456,125	43,792,770	Million 2010 US$	100	100
Agriculture					
Rice	25,997–124,931	53,994–223,967	Million 2010 US$	0.13–0.46	0.12–0.51
Maize	14,988–22,159	26,278–32,889	Million 2010 US$	0.07–0.11	0.06–0.08
Wheat	15,900–61,270	32,707–109,783	Million 2010 US$	0.08–0.30	0.07–0.25
Total	56,886–208,360	112,978–366,639	Million 2010 US$	0.28–1.02	0.26–0.84
Health					
Diarrheal Deaths	1113	824	Million 2010 US$	0.01	0.00
Malaria Deaths	1348	20,748	Million 2010 US$	0.01	0.05
Dengue Deaths	30	499	Million 2010 US$	0.00	0.00
Total	2491	22,072	Million 2010 US$	0.01	0.05
Electricity					
Gas based peak power	33,278	123,183	Million 2010 US$	0.16	0.28
Grand total	**165,966**	**389,980**	**Million 2010 US$**	**0.45–1.19**	**0.59–1.17**

Source: Author's calculations

malnutrition, etc., while we have focused on deaths related to three important diseases: diarrhoea, malaria, and dengue. Impacts on energy infrastructure will be many; the analysis has focused on increased requirement of power generation for meeting peak hour electricity demands. Apart from the three sectors that have been included in the study, climate change impacts hydrological flows, biodiversity, etc., and increased intensity and frequency of extreme events are also critical. Table 5.5 summarizes the cost of inaction calculated across the sectors focused upon in this study.

The analysis highlights some important results. Climate change will result in significant economic losses for India across sectors. Major food crops losses could go up to US$ 208 bn and US$366 bn in 2050 and 2100, respectively (all prices are in 2010 US$). Additional power generation requirements could require additional capital investments of US$33 bn US$ and US$123 bn in 2050 and 2100, respectively, for meeting higher cooling energy needs. Health impacts should be best measured in terms of deaths due to a higher incidence of diseases. Diarrhoeal deaths will decrease with increasing incomes, and deaths due to spread of malaria will increase significantly to 5000 in 2050, and to 19,500 in 2100. Deaths related to dengue will also increase. If disease-related deaths are valued at life time earnings, then loss of economic output will be US$2.5 bn and US$21 bn in 2050 and 2100, respectively. Even with just a few sectors included, and linear representation of the cost of impacts, we arrive at a range of 0.45–1.19% of GDP and 0.59–1.17% of GDP in 2050 and 2100 as the cost of inaction. When nonlinear impacts at higher temperatures are included and other sectors are also valued, the present estimate of cost of inaction is bound to multiply manyfold.

The present analysis, though indicative in nature, is instrumental in giving a good sense of magnitude of the cost of climate change impacts on some key sectors within India. Other methods can be used to understand and evaluate the impacts of climate change for different sectors within India. The analysis intends to provide a solid basis for informed discussions around this issue in India as well as a as a ground for more detailed and insightful study on costs of climate impacts.

References

AEO (2013) Annual Energy Outlook. Energy Information Agency, US Department of Energy, USA

Aggarwal PK (2008) Global climate change and Indian agriculture: impacts, adaptation and mitigation. Indian J Agric Sci 78(10):911–919

Calvin K, Edmonds J, Bond-Lamberty B, Clarke L, Kim SH, Kyle P, Smith SJ, Thomson A, Wise M (2009) 2.6: Limiting climate change to 450 ppm CO_2 equivalent in the 21st century. Energy Economics 31:S107–S120

Chaturvedi V, Clarke L, Edmonds J, Calvin K, Kyle P (2014a) Capital investment requirements for greenhouse gas emissions mitigation in power generation on near term to century time scales and global to regional spatial scales. Energy Economics 46:267–278

Chaturvedi V, Eom J, Clarke L, Shukla PR (2014b) Long term building energy demand for India: disaggregating end use energy services in an integrated assessment modeling framework. Energy Policy 64

Chaturvedi V, Kim S, Smith S, Clarke L, Yuyu Z, Kyle P, Patel P (2013a) Model evaluation and hindcasting: a zero order experiment using an integrated assessment model. Energy. In press

Chaturvedi V, Hejazi M, Edmonds J, Clarke L, Kyle P, Davies E, Wise M (2013b) Climate mitigation policy implications for global irrigation water demand. Mitigation and Adaptation Strategies for Global Change. In Press

Clarke L, Kyle P, Wise M, Calvin K, Edmonds J, Kim S, Placet M, Smith S (2008) CO_2 Emission mitigation and technological advance: an updated analysis of advance technology scenarios. Pacific Northwest National Laboratory Technical Report PNNL-18075; U.S. Department of Energy, Richland, WA, USA

IPCC (2014) Climate change 2014: impacts, adaptation and vulnerability. Fifth Assessment Report, Intergovernmental Panel on Climate Change, Working Group II

Kumar A, Valecha N, Jain T, Dash AP (2007) Burden of Malaria in India: retrospective and prospective view. Am J Trop Med Hyg 77(6):69–78

Kumar SN, Aggarwal PK, Rani S, Jain S, Saxena R, Chauhan N (2011) Impact of climate change on crop productivity in Western Ghats, coastal and northeastern regions of India. Curr Sci 101 (3):332–341

Samra JS, Singh G (2004) Heat wave of March 2004: impact on agriculture. Indian Council of Agriculture Research

Senapati MR, Behera B, Mishra SR (2013) Impact of climate change on Indian agriculture and its mitigating priorities. Am J Environ Protect 1(4):109–111

Shukla PR, Chaturvedi V (2012) Low carbon and clean energy scenarios for India: analysis of targets approach. Energy Econo 34:S487–S495

Stern N (2006) The economics of climate change: the stern review. Cambridge University Press

Wise MA, Calvin KV, Thomson AM, Clarke LE, Bond-Lamberty B, Sands RD, Smith SJ, Janetos AJ, Edmonds JA (2009) The implications of limiting CO_2 concentrations for land use and energy. Science 324:1183–1186

WHO (2014a) Quantitative risk assessment of the affects of climate change on selected causes of death, 2030 s and 2050 s. In: Hales S, Kovats S, Llyod S, Campbell-Lendrum D (eds), World Health Organization

WHO (2014b) Ambient (outdoor) air pollution in cities database

Part II
Energy Challenges and Strategies in Key Sectors

Chapter 6
Low-Carbon Pathways for Urban Development and Mobility in India

Zeba Aziz

Abstract By 2030 India's urban population is projected to almost double, adding another 200 million people to its towns and cities. In the context of a changing climate, this growing urbanization poses a cause for concern for the country's carbon footprint. Cities, in their consumption of energy as electricity or fuel in transportation, are a significant source of carbon emissions, responsible for up to 75% of global carbon emissions. With India's urbanization still at a nascent stage, there is a real opportunity to steer the projected growth away from high-carbon lock-ins and towards a low-carbon pathway. This paper examines opportunities for low-carbon growth in the context of urban development and urban mobility in the country and also illustrates the larger framework within which the aspect of urban mobility is embedded and the multiple pathways available for attaining low-carbon goals in the sector.

6.1 Introduction

India is making an urban transition. Even though the country's urbanization has been low (32%) compared to that of countries like China (45%) and Brazil (87%), this is set to change in the coming years with India's urban population projected to almost double by 2030, adding another 200 million people to its urban areas.[1] In the context of a changing climate, this growing urbanization poses a cause for concern for the country's carbon footprint. Cities, in their consumption of energy as electricity or fuel in transportation, are a significant source of carbon emissions, responsible for up to 75% of global carbon emissions.[2] With India's urbanization

[1]McKinsey Global Institute (2010). India's urban awakening: Building inclusive cities, sustaining economic growth.

[2]UNEP, Cities and Climate Change, retrieved from—http://www.unep.org/resourceefficiency/Policy/ResourceEfficientCities/FocusAreas/CitiesandClimateChange/tabid/101665/Default.aspx.

Z. Aziz (✉)
ICRIER, New Delhi, India
e-mail: zeba.aziz@gmail.com

R. Kathuria et al. (eds.), *Low Carbon Pathways for Growth in India*, India Studies in Business and Economics, https://doi.org/10.1007/978-981-13-0905-2_6

still at a nascent stage, there is a real opportunity to steer the projected growth away from high-carbon lock-ins and towards a low-carbon pathway. In this paper, we examine opportunities for low-carbon growth in the context of urban development and urban mobility in the country.

6.2 GHG Emissions in the Urban Transport Sector

Transportation is a key contributor to carbon emissions on the national and urban scales. In India, the sector accounts for roughly 7% of GHG emissions at the national level, with a 77% growth in emissions between 1994 and 2007, one of the highest across all sectors. This includes both freight and passenger movement through air, rail or road. In 2007, the urban component, primarily limited to the road segment, accounted for roughly 25% of emissions while catering only to 11% of passenger travel.

The urban component of passenger travel is spread over a range of modes, each with a different carbon impact (Fig. 6.1). Among these, private motorized modes are responsible for the highest share of GHG emissions (76%) while serving only 34% of the travel demand,[3] as illustrated. On the other hand, non-motorized modes such as walking and cycling have zero carbon footprints, while catering to a large share of travel demand. Cities aiming for low-carbon futures can use these statistics to take informed decisions regarding infrastructure investment for the city's growth.

As India is urbanizing, the rate of motorization in its cities is also on the rise. The growth in the number of vehicles in tier two cities such as Surat, Pune, Jaipur, etc., is at a rate that is five to six times the rate of their population growth.[4] Nationally too, the share of private modes of transport has increased from 60 to 85% in a period of six decades, with the share of two-wheelers rising from 9 to 72% (1951–2012). Much of this increase can be attributed to the growing gap between the increasing demand for travel and lack of accompanying augmentation in public transport supply, evident from the declining share of public transport at the national level, from 11% in 1951, to 1%, in 2011.[5] The growing motorization has, no doubt, contributed to the high growth in carbon emissions from the sector nationally—the urban per capita contribution to carbon emissions in India is already nine times that of the rural.[6] In addition, it has also had an impact on the urban environment of our cities.[7] In a 2010 study, PM_{10}, an air quality indicator, was found to be in the critical range for 33 of the 49 cities studied. Air pollution from growing traffic is

[3]Planning Commission, Government of India (2011).

[4]Census Bureau, NCAER data, 2006.

[5]Shrivastava et al. (2013).

[6]Parikh et al. (2009).

[7]Shrivastava (see footnote 5)—vehicles contribute close to 60% to air pollution in most large Indian cities.

	Share of Passenger Activity (Passenger-km)	Share of Emissions (MT CO_2)
Air Travel - Inter City	0.36%	15.15%
Rail Travel - Inter City	11.43%	06.06%
Road Travel - Inter City	76.94%	53.03%
Private Motorized - Urban	03.80%	19.70%
Public Transport - Urban	03.32%	06.06%
Non-motorized - Urban	04.16%	0.00%

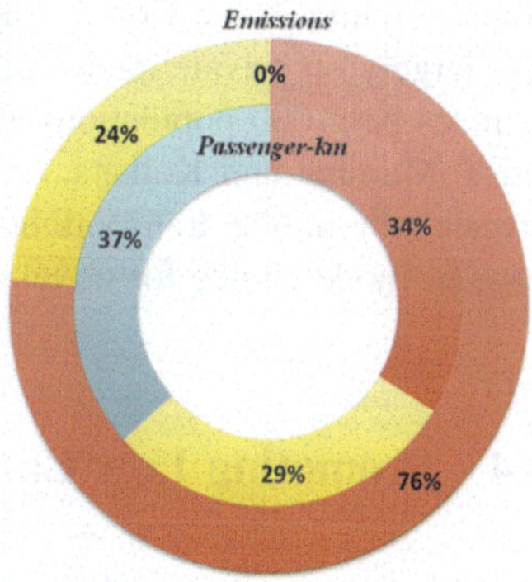

Fig. 6.1 Urban share of carbon emissions in the transport sector. *Source* Low Carbon Strategies —Interim Report 2011, Planning Commission

taking its toll on the health of urban residents. Tackling the issue now thus seems pertinent, not only from the perspective of mitigating carbon dioxide emissions but also from the perspective of productivity or quality of life parameters.

6.3 Urban Growth and GHG Emissions—the Interrelationships

The growth in emissions from urban transport can be tied to several aspects, both from the demand as well as supply sides. The demand for travel and transport is linked to development and growth of the urban economy[8]—greater employment means additional work trips in the city. Spatial growth patterns and distribution of land uses in the city also contribute to the demand for travel—dispersed growth and segregated land uses may increase the need for travel as well as the distances traveled, thus generating a demand for motorized travel.[9] From the perspective of supply, limited coverage by public transit systems or poor amenities for nonmotorized modes encourages the use of private motorized modes. Policies that encourage investment in signal free expressways and wider roads further incentivize the use of automobiles. An aging vehicle stock and poor monitoring of vehicle emissions are aspects that add to the emissions associated with transport.

The fact that there are so many aspects contributing to urban transport emissions helps explain why certain cities such as Ludhiana and Coimbatore, even though small in size, have a much larger per capita carbon footprint than some of the larger cities such as Mumbai and Kolkata.[10] Rapidly growing and prosperous urban

[8]Cameron et al. (2004).

[9]Glaeser et al. (2010).

[10]Reddy et al. (2012).

nodes, Ludhiana and Coimbatore have a nonexistent public transport system and rely largely on private modes for city transportation. Vehicle densities (number of vehicles per 1000 population) in both cities are among the highest in million plus cities. Mumbai and Kolkata, on the other hand, though large, are cities that have invested in public transportation systems that are now well established and a popular mode choice for travel among residents.

6.4 Promoting Low-Carbon Growth in Cities

Considering these multiple and interrelated aspects that affect travel behavior in cities, the latest paradigm for promoting low-carbon growth in the urban sector calls for the Avoid-Shift-Improve approach, a three-pronged approach to dealing with carbon emissions from the transport sector.[11] The "Avoid" segment speaks to the demand side of the sector—advocating for designing cities and economies in a way that minimizes the need for motorized travel. In land use terms, this alludes to cities that are both compact and mixed use. Cities where residents can live, work, shop, and play within a radius efficiently served by zero carbon modes. Of course, pertinent to this is the availability of the relevant infrastructure to support the use of these modes (sidewalks, cycle tracks, etc.). The "Shift" segment recognizes that not all motorized travel will be avoided and advocates for a "shift" in travel modes from low-occupancy, high-carbon private modes, to greener public or nonmotorized modes. Again, pertinent to this is the availability of efficient, affordable, and accessible low-carbon public transport systems that people can "shift" to. The third segment, "Improve" advocates for cleaner vehicles on the road, both public and private, with higher emission and fuel efficiency standards as a final effort to reduce the carbon footprint of the sector. With this segment comes the need for a robust monitoring system to ensure implementation.

The Avoid-Shift-Improve approach provides a comprehensive framework for tackling the issue of urban carbon emissions, recognizing that there is no silver bullet but a range of multi-sectoral interventions that are needed to address the issue. For rapidly urbanizing nations such as ours, it is key that we remain cognizant of these interrelationships while investing in our cities.

6.5 Low-Carbon Pathways in India's Million Plus Cities

At the forefront of India's rapid urbanization are its million plus cities. There are currently 53 such cities in the country, accounting roughly for 43% of the urban population. The remaining 57% of urban population is spread over 7882 cities and

[11]Planning Commission, Government of India (2011).

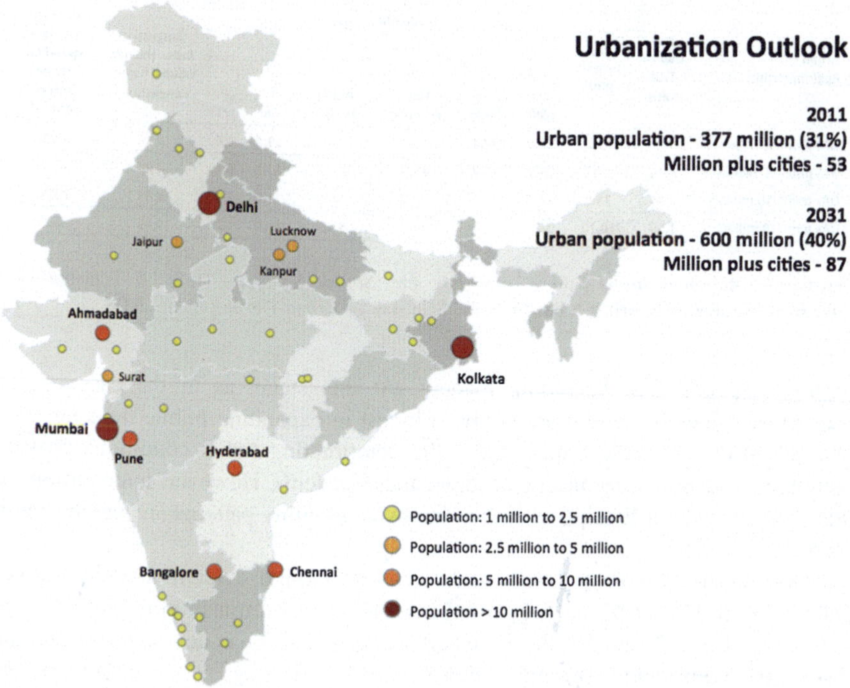

Fig. 6.2 Million plus cities—the hub or urbanization in India. *Source* Census 2011

towns. By 2031, the number of million plus cities is projected to increase to 87, with 34 more cities crossing the million threshold. Growing annually at rates of over 70%, these cities are also at the forefront of carbon emissions, accounting for more than 25% of carbon emissions from the transport sector alone at the national level (Fig. 6.2).[12]

Mobility patterns in million plus cities—Much like with any other system, mobility patterns in Indian cities continue to evolve with city size. To follow the transformation, we map the changes across subgroups of million plus cities, as illustrated in Fig. 6.3. At this scale, one can note the changes with increase in city size. As cities grow, they generate more trips per capita, an indicator of economic growth, as well as longer trips, owing to subsequent urban expansion. With longer trips, deteriorating NMT[13] infrastructure, and increased purchasing power, residents move away from the traditional nonmotorized modes to smaller motorized modes like two-wheelers and three-wheelers and then on to cars. With continued urban growth and increase in travel demand, the volume of traffic in the city rises and with it, the perils of unplanned development such as growing congestion and reduction

[12]Sundar and Ghate (2011).

[13]NMT—Nonmotorized modes of transportation.

Urban Agglomeration	Per Capita Trip Rate	Trip Length (Km)	Mode Choice (%)						Congestion Index (higher value ~ higher congestion)	Average journey speed on major corridors (KMPH)
			Walk	Cycle	Two-Wheeler	Public Transport	Car	IPT		
UAs pop 1-2 million	1.09	4.45	27.8	20.4	23.9	9.0	13.0	6.4	0.19	24.38
UAs pop 2-5 million	1.25	5.43	24.2	18.7	25.0	15.7	10.7	6.0	0.27	22.67
UAs pop 5-10 million	1.41	7.68	22.8	10.0	23.2	25.8	11.2	7.0	0.33	20.00
UAs pop >10 million	1.59	10.70	22.3	9.7	5.3	47.3	10.0	5.7	0.45	16.67

Fig. 6.3 Evolution of travel patterns with city size. *Source* Study on Traffic & Transportation Policies & Strategies in urban areas in India, Ministry of Urban Development, 2008

in travel speeds start becoming evident. At this stage, as in the case of our megacities, cities become more likely to invest in expanding public transport systems to solve the traffic issues, while the deteriorating traffic conditions provide commuters the necessary incentive to use these systems. However, interventions at this stage provide little relief as carbon intensive mobility patterns are already set in place.

Opportunities for low-carbon pathways—It is quite evident that the current patterns of growth or business-as-usual models for urban transport do not speak to low-carbon goals. However, there are certain aspects of urban development in India that do contribute towards those goals. Taking the paradigm of Avoid-Shift-Improve for low-carbon growth as a lens, we locate the opportunities for future work.

Avoid—As discussed earlier, the reduction in travel demand is a first step towards mitigation in the transport sector. Studies have shown that urban form has a significant impact on travel demand, especially the way it shapes the work–home relationship through built density and the distribution of land use.[14] As compared to their western counterparts, Asian cities have traditionally had a high population and built density as well as mixed uses. It is because of these traditional built forms that, despite their comparable population figures, the vehicle miles traveled per capita (VMT) for Asian cities remains as much as three times lower than in their western counterparts.[15] Upcoming urban growth in existing or new cities must preserve and propagate these values as a first step towards low-carbon futures. Of course, this may require a range of interventions such as managing the negative externalities of certain uses/ employment sectors (manufacturing—industries) to allow them to co-exist in cities without harming the urban environment. Certain principles of compact growth, such as Transit-Oriented Development or mixed-use development that is centered on transit nodes (stops for metro, BRT, etc.), are already being tested across cities in India. The urban ministry recently sanctioned a 60% increase in the development capacity (Floor Area Ratio) of areas lying within 500 m of a

[14]Bento et al. (2005).

[15]Litman and Fitzroy (2000).

metro station in Delhi, thus allowing for a higher density of mixed-use development around these transit nodes.[16,17] However, with urban planning being a state subject in India, planning of cities across states may vary vastly based on state polices and local capacities. The ideas for low-carbon growth must therefore be incorporated in the Urban and Regional Development Plans Formulation and Implementation (URDPFI) Guidelines that serve as a planning template for towns and cities across the country.

Shift—Curbing travel demand is only one of the many steps towards low-carbon mobility in cities. Without corresponding availability of well-planned, well-functioning, and affordable infrastructure to support the shift of mass travel from private motorized modes to low-carbon modes of transit, urban density may often result in an increase in congestion and concentration of pollution. Many rapidly growing Indian cities such as Surat and Pune are currently facing these issues.

In terms of carbon mitigation, shifting of mode choice has significant potential and several advances have been made here as well. Studies by the Niti Aayog[18] (erstwhile Planning Commission) and the McKinsey Group[19] indicate an abatement potential of 5–8% from projected emissions over a 15-year period through various scenarios of increasing the share of public transport and NMT in urban India.

The government is making concerted efforts to encourage and facilitate such transitions in growing cities. The Ministry of Urban Development adopted a National Urban Transport Policy in 2006, which for the first time, focused on movement of individuals, rather than of automobiles. The policy has made substantial progress in shifting focus from capacity enhancement projects for automobiles such as construction of new highways and road widenings, to low-carbon mobility options such as improving NMT infrastructure in cities and strengthening public transport systems. The NUTP played a large role in informing investment decisions in the transport segment of the JNNURM[20] program, launched in 2005 to improve infrastructure in growing cities. About 23% of funding in the transport sector was directed towards improvements to public transport in cities. 36 of the 53 million plus cities got funding for new buses while roughly 25 cities got funding for establishing mass transit systems such as BRTS[21] or Metro. Cities such as

[16]Business Standard. 14th July 2015. *Urban development ministry enhances floor area ratio in Delhi by 60%*. Retrieved from http://www.business-standard.com/article/economy-policy/urban-development-ministry-enhances-floor-area-ratio-in-delhi-by-60-115071401088_1.html.

[17]Urban Mobility India, Ministry of Urban Development, Government of India. Transit Oriented Development (**TOD**)- Study for Existing Metro Corridor Between Chattarpur and Arjangarh of Delhi Metro Project of Phase II. Retrieved from http://urbanmobilityindia.in/Upload/Conference/434442dc-9fb7-4722-bd4d-30ead6203541.pdf.

[18]Planning Commission, Government of India (2011).

[19]McKinsey and Company (2009).

[20]Jawaharlal Nehru National Urban Renewal Mission.

[21]Bus Rapid Transit System.

Ahmedabad, Rajkot, Bhopal and Indore have successfully implemented BRT systems thanks to the program.

Several other sector-specific programs have been initiated by the government. The Sustainable Urban Transport Project is one such program launched in 2010 with the support of the Global Environment Facility (GEF), World Bank and UNDP. The project focuses on capacity building of transport professionals in India as well as on the design and delivery of sustainable urban transport systems in five medium-sized, but rapidly growing cities, Indore, Mysore, Hubli-Dharwad, Pimpri-Chinchwad, and Naya Raipur.

Such projects are a move in the right direction for our low-carbon goals. Recent studies show that significant carbon savings are already in place for many of these projects. Ahmedabad's investment in BRTS and in new buses, for example, is already saving the city over 70,000 tons of CO_2 emissions annually. Similarly, improved bus services in Bangalore have saved the city an average of just over 59,000 tons of CO_2 annually since 2005.[22] This is when the mode share of public transport is still well below 50% in both cities. By targeting medium-sized but rapidly growing cities, such programs continue to facilitate capacity enhancements through low-carbon infrastructure that will help limit the urban carbon footprint of future growth as well.

Improve—In addition to initiatives targeting the behavioral aspects of travel choices, (travel frequency, travel distance and mode choice); a key piece in the puzzle of low-carbon mobility is the transportation technology. The Niti Aayog study indicates an abatement potential of up to 17% (from projected emissions over a 15-year period), with improvements in the vehicle and fuel efficiency of transportation modes. However, given the current scenario of transport infrastructure in India, this remains one of the weakest links in the low-carbon story. The issue is threefold here—low quality of existing stock, low standards, and poor monitoring. Recent studies show that more than 25% of on-road vehicles in the country are over 10 years old, some even over 25 years. Emission standards of Euro 4 (Bharat Stage IV) for four-wheel vehicles are currently applicable in only 14 cities. In terms of fuel efficiency, India is the only automobile manufacturing country that does not have efficiency standards in place yet. The Bureau of Energy Efficiency will be introducing norms for private vehicles, but only by 2017.[23] In addition, the emission monitoring systems (PUC) are lax. Even in the capital city of Delhi, only 21% of vehicles show up for emissions testing despite the provision of heavy penalties.[24] Significant work will thus be needed to achieve the abatement potential under this option. With new systems of public transit being introduced in cities as part of urban regeneration efforts, it is important that high-carbon lock-ins are avoided

[22]Prabhu and Pai (2012).

[23]The New Indian Express. 17th January 2014. *BEE Forms New Fuel Efficiency Standards*. Retrieved from—http://www.newindianexpress.com/nation/BEE-Forms-New-Fuel-Efficiency-Standards/2014/01/17/article2003990.ece.

[24]Lakshmi et al. (2014).

through access to better technology. But conflicts may arise. A recent study noted that the technology for new buses mandated as part of the Jawaharlal Nehru National Urban Renewal Mission (BS-IV), while meeting higher emission standards (less air pollution) was less fuel-efficient than the preceding standards.[25] A range of similar aspects of capacity, access, and affordability factor in the decisions as well.

6.6 Conclusion

This paper illustrates the larger framework within which the aspect of urban mobility is embedded and the multiple pathways available for attaining low-carbon goals in the sector. Given the interrelatedness between urban form, land use, and mobility patterns, it is imperative that our rapidly growing cities adopt a comprehensive view while looking at reducing their carbon footprints, right from siting of major employment centers and other key land uses, to the distribution of amenities and investment in infrastructure. At the national level too, the carbon impact of urban policies needs to be taken into account while rolling out sector-specific reforms, especially in the case of urban programs such as JNNURM or the newly launched AMRUT[26] and the Smart Cities initiative[27] by the Ministry of Urban Development.

References

Bento AM, Cropper ML, Mobarak AM, Vinha K (2005) The effects of urban spatial structure on travel demand in the United States. Rev Econ Stat 87(3):466–478

Cameron I, Lyons TJ, Kenworthy JR (2004) Trends in vehicle kilometres of travel in world cities, 1960–1990: underlying drivers and policy responses. Transp Policy 11(3):287–298

Fok T, Prabhu A, Bachu P (2013) The high cost of low emissions standards for 2 bus-based public transport operators in India: 3 evidence from Bangalore. Retrieved from docs.trb.org/prp/14-0612.pdf

Glaeser EL, Kahn ME (2010) The greenness of cities: carbon dioxide emissions and urban development. J Urban Econom 67(3):404–418

Lakshmi CS, Sharma S, Sundar S, German J, Bansal G, Walsh MP (2014) Establishing a national in-use vehicle testing programme in India. Shakti Foundation

Litman T, Fitzroy S (2000) Safe travels. Victoria Transport Policy Institute, www.vtpi.org. Maunder, DAC, and Pearce, TC

McKinsey & Company (2009) Environmental and energy sustainability: an approach for India

---

[25]Fok et al. (2013).

[26]Atal Mission for Rejuvenation and Urban Transformation (AMRUT), Ministry of Urban Development, Government of India.

[27]Smart Cities Mission, Ministry of Urban Development, Government of India. *Smart City Features*. Retrieved from http://smartcities.gov.in/writereaddata/Smart%20City%20Features.pdf.

Parikh J, Panda M, Ganesh-Kumar A, Singh V (2009) CO_2 emissions structure of Indian economy. Energy 34(8):1024–1031

Planning Commission, Government of India (2011) Interim report of the expert group on low carbon strategies for inclusive growth

Prabhu A, Pai M (2012) Buses as low-carbon mobility solutions for urban India: evidence from two cities. Transp Res Record J Transp Res Board 2317:15–23

Reddy BS, Balachandra P (2012) Dynamics of urban mobility: a comparative analysis of megacities of India (details of publication?)

Shrivastava RK, Neeta S, Geeta G (2013) Air pollution due to road transportation in India: a review on assessment and reduction strategies. Environ Res Develop 8(3)

Sundar S, Ghate AT (2011) Transport and energy: the Indian perspective. In Transport moving to climate intelligence. Springer, New York, pp 147–159

Chapter 7
Strategies to Lower Carbon Emissions in Industry

Saon Ray and Nandini Kumar

Abstract Major anthropogenic sources of these emissions include vehicular transport and industrial activity. The industrial sector accounts for the largest share of delivered energy consumption and is expected to consume over half of the global delivered energy in 2040. 31% of the world energy consumption or 200 quadrillion British Thermal unit (Btu) is consumed in the industrial sector worldwide. This is expected to rise to 307 Btu in 2040. Commensurate with the energy consumption in the world, worldwide energy-related carbon dioxide emissions are expected to increase from 31 billion metric tons to 36 billion metric tons in 2020 and 45 billion metric tons in 2040. Greenhouse gas emissions for 2012 are estimated to be 31.6 Gt (Gt). This paper examines how much carbon dioxide can be attributed to industry and what are the ways to decarbonize industry in the Indian context. While the regulatory mechanism required to reduce energy intensity is in place for the most energy-intensive industrial sectors, the generation of electricity is still primarily dependent of coal. Use of renewable needs to be stepped up and is line with India's commitment of INDCs.

7.1 Introduction

The world energy consumption is expected to grow by 56% between 2010 and 2040, a figure based on projections of the *International Energy Outlook* 2013 (US Energy Information Administration 2013). The industrial sector accounts for the largest share of delivered energy consumption[1] and is expected to consume over half of the global delivered energy in 2040. 31% of the world energy consumption,

[1]Delivered energy is measured as the heat content of energy at the site of use.

S. Ray (✉)
Indian Council for Research on International Economic Relations, New Delhi, Delhi, India
e-mail: sray@icrier.res.in

N. Kumar
CII-ITC Centre of Excellence for Sustainable Development, New Delhi, India

R. Kathuria et al. (eds.), *Low Carbon Pathways for Growth in India*, India Studies in Business and Economics, https://doi.org/10.1007/978-981-13-0905-2_7

or 200 quadrillion British Thermal units (Btu) are consumed in the industrial sector worldwide. This is expected to rise to 307 Btu in 2040. Commensurate with the energy consumption in the world, worldwide energy-related carbon dioxide emissions are expected to increase from 31 billion metric tons to 36 billion metric tons in 2020, and 45 billion metric tons in 2040.

Major anthropogenic sources of these emissions include vehicular transport and industrial activity. When vehicles burn fossil fuels, carbon compounds of which the fuel is made, combine with oxygen in air aided by high temperatures within the engine, to form carbon dioxide, a greenhouse gas. The reaction is inevitable, and strategies to decarbonize industry must work around this fact. Together with carbon dioxide, other gases are emitted when fossil fuels burn: the most prominent of them are the oxides of nitrogen, known to be pollutants. Thus, whenever fossil fuels are burnt, both, greenhouse gases and pollutants are released to the atmosphere. Greenhouse gas emissions for 2012 are estimated to be 31.6 Gt (Gt) (International Energy Agency 2013).

The Fifth Assessment report (FAR) (2014)[2] of the IPCC notes that more than half the global population is urban and the kinds of towns and cities that emerge in the future will be critical in determining energy use and carbon emissions. Rural populations generally have lower per capita energy consumptions compared to urban areas in developing countries (IEA 2008).[3] Urban areas act as engines of economic growth and activities, and an assessment by the World Energy Outlook 2008 shows that 19.8 Gt or 71% of CO_2 emissions occur in urban areas accounting for 222–330 EJ of primary energy.

India and China have emerged as important consumers of energy and their energy intensity is yet to peak. The pathways taken by the India and China in the future will determine the total energy consumption in the world and the carbon dioxide emissions. These in turn depend on the demand–supply balance of fuels like oil, coal, renewable energy use, etc. and the price of each. How much of carbon dioxide can be attributed to industry and what are the ways to decarbonize industry? This is the central question that this paper will try to answer in the context of India. Comparisons with China are inevitable and hence will be made throughout the paper.

The paper is organized in the following way: Section 7.2 discusses the consumption of energy worldwide by industry with an aim to identify the energy-intensive industrial sectors and the energy consumed by them. The impact of carbon emissions on climate are outlined in Sect. 7.3. Section 7.4 sketches the potential for savings of carbon emissions in different industries through improvement in energy efficiency. Section 7.5 presents the Indian case: consumption of energy and the fuel mix by sector, and emissions from energy-intensive industries. Since much of industries in the Indian context are located in and around cities (Guttikunda and Jawahar 2012), this has also been discussed separately. The policy

---

[2]Seto et al. (2014).

[3]De industrialization.

framework needed to achieve a low-carbon pathway has been identified in Sect. 7.6. Section 7.7 concludes.

7.2 Consumption of Energy by Industry

The energy sector accounts for two-thirds of greenhouse gas emissions, as more than 80% of the global energy consumption is based on fossil fuels. Non-OECD countries accounted for 60% of the global emissions of CO_2 in 2012, up from 45%, in 2000.

According to the International Energy Outlook 2013, the industrial sector uses more delivered energy than any other end-use sector. There are many diverse sectors in the industrial sector, from manufacturing to mining to construction. Each of these sectors has different energy requirements and emits different greenhouse gases.[4] Energy is used for processing, assembly, producing steam, cogeneration, etc. Various fuels including natural gas, petroleum and coal are used by the industrial sector (International Energy Outlook 2013).

As mentioned earlier, depending upon the material/mineral/metal being processed in a particular factory/industry, different gases and particles are released at the high temperatures typically encountered in manufacturing processes. More often than not, other polluting gases are also released in addition to carbon dioxide. Some examples are sulfur dioxide, hydrogen chloride, nitric acid, fluorine, and particles such as soot.[5]

The following figure shows the energy consumed by industry in OECD and non-OECD countries, and the world, in 2010, by energy source. As we can see from the figure, the largest consumption by industry is petroleum, followed by coal and then natural gas. For the OECD countries, the largest consumption is petroleum followed by natural gas and then coal. This trend is reversed in the non-OECD countries, with coal dominating the other sources. The other important point to be noted is that consumption in the OECD countries was 71.9 quadrillion Btu, while that in the non-OECD countries was 128.1 quadrillion Btu in 2010. It is projected that this will increase to 87.1 and 219.8 quadrillion Btu, respectively, in 2040. Coal consumption is projected to increase from 45.5 quadrillion Btu for the world in 2010 to 79.6 quadrillion Btu in 2040 of which 70.4 quadrillion Btu will be consumed in the non-OECD countries and the rest in the OECD countries. Petroleum and other liquids (natural gas liquids) are expected to rise from the present consumption of 57.2 quadrillion Btu in 2010, to 78.2 quadrillion Btu in 2040, with the

[4] CO_2 emissions differ by fuel and the use of fuel is not distributed evenly across industrial subsectors.

[5] Ozone precursors (hydrocarbons and nitrogen dioxide) are also emitted during combustion of fossil fuels and associated with transport and industrial activity. Ground-level ozone is formed by reaction of sunlight on air containing these precursors. Ozone at ground level is not only a pollutant whose levels in urban areas worldwide are rising but is also a greenhouse gas.

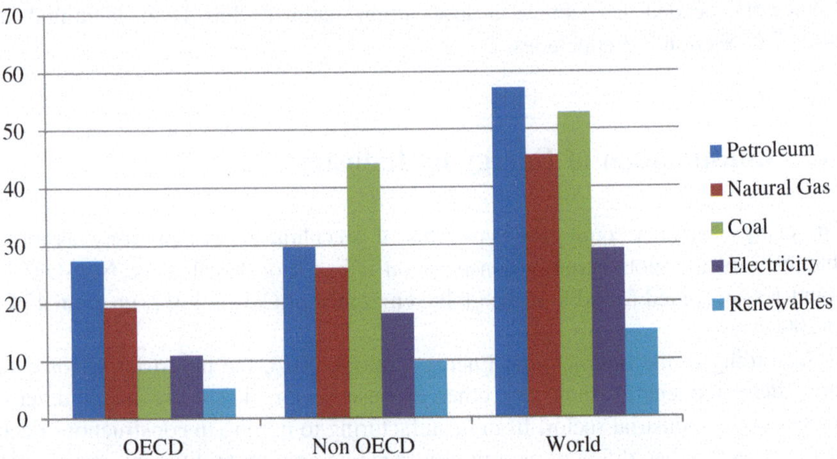

Fig. 7.1 Energy consumption by industry (quadrillion Btu), 2010. *Source* Compiled from information in International Energy Outlook 2013

share of the non-OCED countries rising from the present 29.8 quadrillion Btu to 46.5 quadrillion Btu in 2040 (Fig. 7.1).

Estimates for India and China show that while China's industry consumed 8.4 quadrillion Btu in 2010, Indian industry consumed 3.2 quadrillion Btu (International Energy Outlook 2014) of petroleum and other liquids. In 2040, it is expected that China's consumption will rise to 12.2 quadrillion Btu while that of India's will rise to 5.7 quadrillion Btu.

China's coal consumption in 2010 was 69 quadrillion Btu and used primarily for electric power (50%) followed by industry (45% in 2010). Coal consumption is shifting from industry to the electric power sector and in 2040, and the share of coal in industry is expected to be 41% while that of the power sector 57%. Total consumption of coal in China is expected to be around 125 quadrillion Btu in 2040. In India, coal consumption was 12.6 quadrillion Btu in 2010 and is expected to rise to 22.4 quadrillion Btu in 2040. Coal fuelled 68% of India's total electricity generation in 2010, while industries used 32%. Coal consumption for electricity generation is expected to rise from 8.2 quadrillion Btu to 15.6 quadrillion BTU in 2040 (International Energy Outlook 2013).

The chemicals, pulp and paper, iron and steel, refining and the non-metallic minerals industries account for one half of the all energy used in the industrial sectors. Moreover, these industries also emit large volumes of carbon dioxide due to both, combustion and production processes.

7.3 Emission by Industry—Effect on Climate

At the UNFCCC Conference of Parties in Cancun in 2010, governments agreed to limit the rise in average global temperatures to less than 2 °C, and in order to maintain this temperature, to reduce greenhouse gases. This translates to limiting the atmospheric concentration of greenhouse gases to below 450 parts per million (ppm) of $GtCO_2e$). However, global energy-related CO_2 emissions rose to 31.6 Gt (Gt) in 2012 (International Energy Agency 2013b), an increase of 0.4 Gt over the 2011 levels. CO_2 levels in the atmosphere reached 400 ppm in May 2013, with a jump of 2.7 ppm in 2012.

The IPCC Fifth Assessment Report (2014) has estimated that the maximum amount of carbon dioxide that could be emitted over time while staying within the 2 °C global warming limit is 3670 $GtCO_2e$[6] of which 1900 $GtCO_2e$ has been emitted since the nineteenth century. This leaves 1000 $GtCO_2e$ to emit in the future.

Greenhouse gas emissions for 2010 are estimated to be 50.1 Gt (Gt) of carbon dioxide equivalent ($GtCO_2e$) per year (UNEP 2013). This is 14% higher than the median estimate of the emission level in 2020 and hence affects the chance of meeting the 2 °C target by that year. The World Bank (2012) has discussed the effect of 4 °C rise in mean temperature particularly from the point of view of developing countries. It argues that there is a 20% likelihood of exceeding the 4 °C by 2100. This, if it does occur will have an effect on several climate indicators, among which is included the emissions scenario.

The World Bank (2012) projects that even if the emissions pledge made at Copenhagen and Cancun are fully met, the world will be placed on a trajectory of well over 3 °C. Alternatively, if the pledges are not met the likelihood of exceeding the 4 °C are 20%.

The relative contribution to global emissions has changed, with the share of developed countries rising from 48.2 to 59.1% in the period 2000–10. Carbon dioxide emitted by fossil fuel combustion and cement production contributes the most to greenhouse gas emissions; this grew by 2.0% in 2013 (UNEP 2014).

Considering these dire consequences of climate change projected above, it is imperative to search for low-carbon pathways for countries. In the context of existing industries, one of the simplest ways of decarbonization is by improving energy efficiency over all processes.

7.4 The Role of Energy Efficiency

The role of energy efficiency measures for limiting warming to below 2 °C has been discussed in the Banerjee et al. (2012) and the UNEP Emissions Gap report (2014). The energy intensity improved, on an average, by 1.6% globally annually between

[6]Assuming that carbon dioxide is the only anthropogenically generated greenhouse gas.

2002 and 12. For the OECD countries, the cumulative energy savings was 1731 million tons of oil equivalent (Mtoe) over the period 2001–11. Because industrial activity, at present, depends almost totally on fossil fuels, any move to improve the efficiency of machines run or burning of fuels will reduce emissions of carbon dioxide and also the other pollutants associated with that industry. Making electrical devices/machines/equipment more efficient will mean lower demands on power plants which produce the electric power, resulting in multiple benefits to the environment (IEA 2014).

IEA (2007) estimates that the manufacturing industry can improve its energy efficiency by 18–26%, while reducing its CO_2 emissions by 19–32%. These improvement options could contribute between 7.4 and 12.4% reduction in global energy and process-related carbon dioxide emissions. This includes savings from chemicals and petrochemicals (5–6.5 Exajoule per year (EJ/year), iron and steel (2.3–4.5 EJ/year), cement (2.5–3 EJ/year), and paper and pulp (1.3–1.5 EJ/year). The study sets out a new set of indicators for country-level energy level efficiency levels based on intensive consultations and historic trends and current efficiencies. The study also estimates the savings in carbon dioxide emissions per year—the highest is in the cement sector with 480–520 Mt CO_2/year.

The largest industrial sector consumer of delivered energy is the chemicals sector, followed by the iron and steel and nonmetallic minerals sectors. The chemicals and petrochemical industry accounts for more than 30% of the global industrial energy use: the bulk of this is concentrated in the feedstock used in the petrochemical sector. The chemicals sector accounted for 16% of direct CO_2. This cannot be reduced through energy efficiency measures (IEA 2007). China and India's chemicals and petrochemicals industries accounted for 20% of industrial energy use, by region, in 2004. The sector produces a large number of products. The age of a plant often defines its energy efficiency with older plants being generally less energy efficient. Improved final energy potential in this industry is 8.5–11 EJ/year. This includes 4 EJ of fuel savings potential, while the remainder is through electricity savings, combined heat and power (CHP), recycling and energy recovery (IEA 2007). A report by CII (2013) estimates the Specific Energy Consumption of the organic and inorganic chemicals sector in India for 2011–12. The energy reduction potential is estimated to be around 1–20% for the sector based on energy audits conducted by CII.

A special feature of the cement industry is the fact that limestone (calcium carbonate) is a raw material used in significant quantities; when limestone is heated, it emits carbon dioxide because of the carbonate it is composed of. This is in addition to any carbon dioxide that might result from fossil fuels being burnt in the cement plant. Certain products are therefore, associated with more greenhouse emissions because of the raw materials used in their manufacture. In these cases, searching for substitutes through scientific research could provide a solution.

7.5 The Case of India

Much of the growth in energy consumption is expected from countries like China and India and will be driven by strong long-term economic growth there (International Energy Outlook 2013). China and India have been among the world's fastest growing economies for the past two decades, growing at an average of 10.4 and 6.4%, respectively. Even though real GDP growth in the two countries has been slower since then, and was 7.2% for China and 5.5% for India in 2012, the economic recovery from the recession has been led by these countries. The strong economic growth in the two countries is expected to proportionately increase their energy demand (IEA 2008). Since 1990, energy consumption in both countries accounted for 10% of the total world energy consumption in 1990, and 24% in 2010.

7.5.1 *Energy Consumption and Fuel Mix by Sector*

India consumed 3.1 million barrels of liquids per day in 2009, and 3.3 million barrels of liquids per day in 2010. This is expected to rise to 6.1 million barrels per day in 2040 (International Energy Outlook 2014). The following graph shows the actual and projected consumption of world petroleum and other liquids by various sectors in India from 2010 to 2040 (Fig. 7.2).

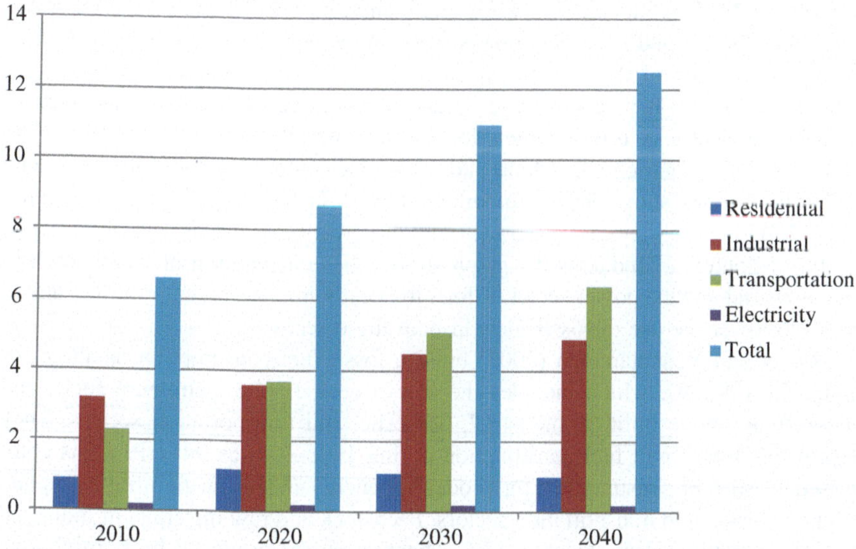

Fig. 7.2 Consumption of world petroleum and other liquids by source (quadrillion Btu) by India, 2010–2040. *Source* Compiled from information in International Energy Outlook 2014

There are three strands of the literature dealing with economic growth, energy consumption and environmental pollutants. The first strand of thought deals with the environment-pollutant output nexus and is closely associated with the empirical testing of the Environmental Kuznets curve (EKC). The findings in this strand of literature are not conclusive: higher national income does not lead to greater effort in combating pollutants. The second strand of literature deals with the relationship between income and energy consumption. The main conclusion of this research is that the findings vary from country to country and even the direction of causality runs from either income to energy consumption or vice versa or both. The third strand of thought examines the dynamic relationship between income, energy consumption, and environmental pollutants. Ajmi et al. (2015) use data for G 7 countries to show that this relationship changes over time, for example, the results show a bidirectional time-varying causality between energy consumption and CO_2 emissions in the case of the United States.

In the case of India, Ghosh (2010) finds bidirectional causality between carbon dioxide emissions and economic growth, and a unidirectional relationship between economic growth and energy supply, for the period 1971–2006. The author feels that energy conservation and energy efficiency measures can be implemented to minimize the wastage of energy, as such measures would narrow energy demand–supply gap. However, the absence of long-run causality between carbon emissions and economic growth implies that in the long-run, the focus should be on harnessing energy from clean sources to curb carbon emissions, which would not affect the country's economic growth. Some of these pathways have been suggested by Gupta, in this volume.

India is among the top ten countries in the world in Total Primary Energy Supply (TPES) in 2013, producing 775 Mtoe and a share of 6%. China, on the other has overtaken the US and is number one position in the world with production of 3022 Mtoe and a share of 22% in the world in 2013. The consumption of energy by China and India was respectively at 22 and 6% in 2013 (IEA 2015). The share of coal in the electricity mix is largest for Asia among all the regions in 2013. Coal provided 76% electricity in China and 73% of electricity in India in 2013. The following graphs show the use of different fuels in the various sectors of India (Fig. 7.3).

In the context of India, as the graph shows, the consumption of coal is high in the industrial and the power sector. Efforts to decarbonize the country must examine ways to lower the use of fossil fuels in both these sectors.

Reddy and Venkataraman (2002) build a fossil fuel consumption database for India for 1996–97, which includes the use of coal, lignite, petroleum fuels, and natural gas use in utilities, industrial, domestic, and transportation sectors. They report that total fossil fuel consumption during 1996–97 was 9411 PJ. This comprised fossil fuel consumption for producing energy (83%) in the utilities, industrial, domestic, and transportation sectors. Feedstock and raw materials in industrial processes comprised the balance 17%. The energy was produced by burning coal (54%), diesel (18%), natural gas (12%), and other fuels (16%). Of the 9411 PJ, the highest sectoral consumption came from the industrial sector (43%), followed by

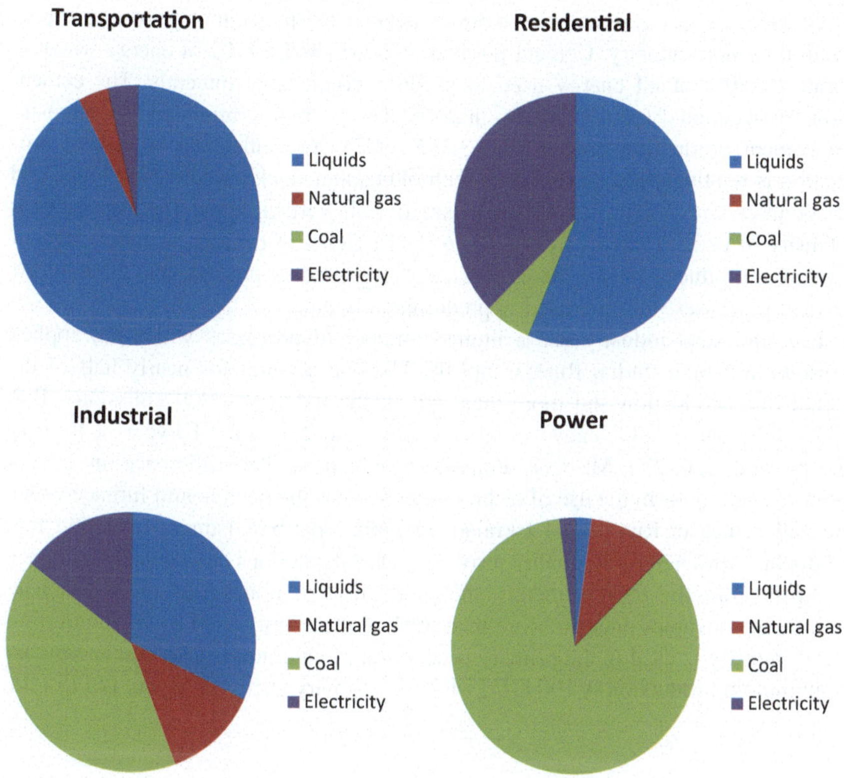

Fig. 7.3 Fuel mix of different sectors in India. *Source* Compiled from information in International Energy Outlook 2013

utilities (37%), transportation (17%), and domestic (3%). The breakup of industries in the total consumption of energy was iron and steel (10%), cement (2%), fertilizers (8%), brick kilns (3%), and other industries (19%).

Sathaye et al. (2005) review the energy efficiency of five energy-intensive sectors —fertilizers, textiles, chlor alkali, cement, and petroleum refining in India. They observe that in each of the sectors, the industry has taken steps to reduce energy intensity. Sometimes the improvement has come due to stricter environmental norms as in the case of chlor alkali and in other instances the change in the fuel as in the case of fertilizers. UNIDO (2010) has examined the energy use of manufacturing industry as a benchmarking exercise worldwide. The percent of improvement for India is as high as 27% in the steam cracking process. A number of industry sectors such as ceramics, textiles, leather, foundry, and other processed metal are dominated by SMEs which are largely energy inefficient. Most benchmarking exercises fail to captures this, but this needs to be recognized as well. Any attempt to improve energy efficiency in industries without improving the energy efficiency of SMEs will be counterproductive.

Greenhouse gas emissions by industry depend on the technology used for production by that industry. Cement production accounts for 8 EJ of energy which is about 70–80% of all energy used to produce nonmetallic minerals. The cement industry accounted for 1.8 Gt CO_2 in 2005. The average primary energy intensity for cement production ranges from 3.4–5.3 GJ/t. The efficiency of cement production is relatively low in countries with old capital stock based on wet kilns, and in countries with a significant share of small-scale vertical kilns. The average CO_2 intensity of cement ranges from 0.65 to 0.92 t CO_2/t of cement. In India, 50% of cement is produced by the dry process, 25% by the wet process, and 16% by the vertical processes: the fuel used is predominantly coal.

Iron and steel industry use a limited number of processes which are applied worldwide. China, India, Russia, and the Ukraine account for nearly half of the global iron production and more than half of the industry's CO_2 emissions (IEA 2007). The energy efficiency potential varies from 2.3 to 2.9 EJ/year of primary energy and 220–270 Mt CO_2 emissions reduction. The difference in energy intensities come from the use of technologies such as the open hearth furnace which are still in use in Russia and Ukraine. In India, coal-based direct reduction iron production which use low-quality resources has a deleterious environmental impact.

Apart from the ferrous metals, the production of nonferrous metals such as aluminum consumes energy. More than 50% of the energy used by the nonferrous metal industry is used in the primary production of aluminum. The final energy use in aluminum production is 100 GJ/t, while the primary energy intensity is 175 GJ/t.

7.5.2 *Cities and Emissions*[7]

As more and more people live in urban settings now, they are likely to have a significant impact on GHG emissions. The Global Energy Assessment puts urban energy use between 180 and 250 EJ, which corresponds to an urban share ranging from 56 to 78% of global final energy use. Grubler et al. (2012) estimate global urban energy-related CO_2 emissions of 8.8–14.3 Gt, which correspond to between 53 and 87% of CO_2 emissions from global final energy use. The large variation in the emissions arises from problems in defining boundaries and delineating urban areas, unavailability of data on GHG emissions for urban and rural areas, and the approach (bottom up or top down) used in estimating emission shares.

Although rates of urbanization growth vary across countries, 55% of global urban expansion is expected in China and India in 2030 (Seto et al. 2012). This has

[7]Industries tend to be located in or near cities. While the Ministry of Environment and Forests note that the site of industries has to be 25 km away from cities and the spatial direction of growth for a decade must be assessed while taking a decision regarding the site of an industry, Lall and Mengistae (2005) observe that local business environment and agglomeration economies significantly influence the business location choices across cities in India.

implications for CO_2 emissions as well in these countries. O' Neill et al. (2012), used the integrated Population Economy Technology Science (iPETS) model and extrapolated the urbanization pattern for India and China till 2050. They found that urbanization has a less than proportional effect on emissions and energy use in case of China and India. In the case of India where effect is smaller than in China, they conclude that this is due to differences in the current rate of urbanization and urban–rural disparity in income. This suggests that a shift in India towards urbanization associated with greater industrialization in urban areas could lead to a significantly different effect of urbanization on emissions and energy use.

Ramachandra et al. (2015) estimate the amounts of three greenhouse gases, carbon dioxide (CO_2), methane (CH_4), and nitrous oxide (N_2O), in seven Indian cities (Delhi, Greater Mumbai, Kolkata, Chennai, Greater Bangalore, Hyderabad, and Ahmedabad) for 2009. The GHG footprint of the cities ranged from 9124 Gg CO_2 equivalent to 38,633 Gg CO_2 for Delhi. The paper analyzes the major contributor to the GHG emissions and finds that while the transportation sector is the largest, industry contributes 22% of carbon dioxide equivalent emissions in Ahmedabad, while in the case of Chennai and Kolkata, these figures are 20 and 17%, respectively. For the other cities, it varies from 8% for Greater Mumbai and Delhi, 11% for Hyderabad, and 12% for Greater Bangalore.

Using an air pollution production function for Indian cities, Holian (2014) tests for effect of income, literacy, and population on four measures of air pollution and finds that the relationship between social and economic development and pollution varies across pollution type. While the dependent variable in the estimated equation is measures of pollution type: average SO_2, average NO_2, average respirable particulate matter (RSPM) and average suspended particulate matter (SPM), the independent variables included income (at the district level), total output in manufacturing divided by total output in all categories in the district in which the city is located, and literacy at the city level. The results obtained indicate that there is a negative relationship between income and particulate matter and a positive relationship between income and nitrogen dioxide. Further, he finds strong evidence that cities with a higher concentration of manufacturing have higher levels of all four measures of pollution. Relating the level of economic activity with 18 industry categories, he finds that while most categories are positively related with SO_2, output was negatively correlated with particulate matter levels. The highest positive correlation with NO_2 is with storage, public administration. This may be explained by the fact that (since NO_2 is largely attributable to vehicular transportation (CPCB)) many residents of the cities commute in motorized vehicles.

A particular feature of cities, especially those in India, is that some factories have continued to operate or have even been set up within the urban area, so that in addition to vehicular transport as a source of greenhouse gases (primarily carbon dioxide), there are sources of other pollutants such as sulfur dioxide. Guttikunda and Jawahar (2012) present results from application of the SIM–air modeling for six Indian cities. They note that industries are within city limits in all six cities. They present total emissions by source for the six cities (Pune, Chennai, Indore, Ahmedabad, Surat, and Rajkot) in 2010, and also an emissions inventory of PM_{10},

$PM_{2.5}$, NO_x, SO_2, CO, and CO_2 for the cities. The results indicate that CO_2 emissions vary from 7.4 million tons in Rajkot to 31.6 million tons in Chennai for 2010. They apportion the sources of PM and find that transport is the dominant source while industries, including brick kilns, are significant contributors to PM in all the cities (highest in Rajkot).

If a marker of pollution for industry has to be identified, then within a city, sulfur dioxide might be a suitable candidate (vehicles give off little sulfur dioxide in their emissions). Guttikunda and Jawahar (2012) report highest concentrations of SO_2 from Chennai (18,100 tons) and Ahmedabad (15,100) tons, while for the other cities, it varies from 2200 tons for Rajkot to 4100 tons from Pune.[8] Ahmedabad is known as an industrial city but Chennai has power plants within city limits (as does Ahmedabad). This also brings to the fore that the kind of industry (or power plants) also matters in determining emissions.

Reddy and Venkatarman (2002) estimated that SO_2 emissions from fossil fuel combustion in India from 1996 to 97 were 4.03 Tg SO_2 per year. They estimated that 756 large point sources (LPS) accounted for 62% of total emissions. SO_2 from utilities were mainly due to combustion of coal which contributed 43% to SO_2 emissions. $PM_{2.5}$ emissions from fossil fuel combustion were largely due to use of coal in power plants, which were responsible for 79% of such emissions, followed by brick kilns (8%) and diesel in transportation (7%). Black carbon (BC) and organic matter (OM) were 29 and 22%, respectively, of total emissions. Diesel use in transportation was the single largest contributor to BC, accounting for 58%. Brick kilns fired with coal resulted in 48% of OM emissions.

7.6 Policy Framework for Low-Carbon Pathways

The pathways taken by India and China in the future will determine the total energy consumption in the world and the carbon dioxide emissions. These in turn depend on the demand–supply balance of fuels like oil, coal, renewable energy use, etc. and the price of each. Technological advancements would also play an important role as will the availability of newer sources of energy such as shale. Achieving the target of 2 °C would require higher-than-current emission reduction rates, improvements in energy efficiency, and introduction of zero-to-low-carbon technologies, at rates faster than have been historically experienced (UNEP 2013).

Indian consumption of liquids was 35% of China's 9.3 million barrels per day in 2012, and is expected to be the second largest growth in the world between 2010 and 2040 (at 5 million barrels per day). In India, petroleum consumption is heavily oriented towards diesel, which represented 42% of product volume in 2012. Diesel

[8]Guttikunda and Jawahar (2012) report that in all six cities, industries are within city limits and in areas of high population density.

is used in transportation, irrigation, manufacturing, and electricity generation (International Energy Outlook 2013).

India is the third largest producer and consumer of coal in the world in 2010 and coal production is projected to increase at 1.6% annually from 612 million tons in 2010 to 1 billion tons in 2040. India consumes almost all its coal and imports coal, which from 140 million tons in 2011 is expected to increase to 300 million tons in 2040. India's coal consumption in 2010 was 12.6 quadrillion British thermal units (Btu), 68% of which fuelled electricity generation while the balance of 32% was used in industry to produce iron and steel, cement, bricks, and other materials (International Energy Outlook 2013).

Natural gas accounted for 10% of India's overall energy consumption in 2010 and its growth is expected to be lower than that of the other fuels in the period 2010–2040 (International Energy Outlook 2013). India imports part of its natural gas requirement and has less than 1% of proven reserves of natural gas in the world (Ray et al. 2014). The share of nuclear and renewables like wind and solar is low at present in India, but is expected to increase with plans by the government to promote these sectors. India's dependence on coal could decline, depending on which of these sectors can cater to India's growing energy demand. However, the technologies required for achieving negative emissions in the energy and industrial sector have not been deployed at significantly large scales (UNEP 2013). Hence, several possible pathways are possible for India's low-carbon transition and some of the possible pathways that India could embark upon need be discussed in the context of specific industries.

The low-carbon pathway for India will be determined by the regulatory framework to achieve energy efficiency in industry and also by the technology available in the industries to switch to lower carbon options (IEA 2014).

Industrial energy efficiency policy, in order to be successful, must have four elements all of which are part of the India's policy. The regulatory structure under the aegis of the National Action Plan on Climate Change (NAPCC) in India has these four main features. First, energy audits in the nine most polluting sectors are mandatory since 2007. Second, from 2006, energy labeling schemes for appliances have been introduced by the Bureau of Energy Efficiency (BEE). The minimum efficiency standards have been set through the National Mission for Enhanced Energy Efficiency in Industry which is part of the NAPCC. Energy efficiency targets were set for fifteen industries which were identified for the Perform, Achieve and Trade (PAT)[9] Scheme to reduce their energy intensity. Finally, a guide to the energy management systems standards that are to be developed is provided by the NAPCC. This helps in achieving energy management as part of the company wide energy policy.

Coming to the energy intensity of Indian industries, the Indian Renewable Energy Development Agency (IREDA) along with the Confederation of Indian Industries (CII) have identified the 20 most energy-intensive industries and outlined

[9]Press release: http://pib.nic.in/newsite/mbErel.aspx?relid=85182.

the processes that can be undertaken by these industries to increase energy efficiency in 16 of these industries. Of these 16 industries, several are also identified as most polluting by the NAPCC and some were included in the PAT scheme.

7.7 Conclusion

Scientifically, the reason decarbonizing is necessary lies in the concept of a source and sink and cycling between them, for all components of the land, water, and air. Carbon dioxide from any source accumulates in the atmosphere and reacts with other components there to form other chemical compounds. This conversion to sinks (in which carbon dioxide is locked up) is governed by reaction *rates* which are not alterable and so, some reactions can take months, years, and even decades. Therefore, to prevent accumulation—and resultant enhancing of the greenhouse effect by carbon dioxide—it becomes necessary to cut down its sources, that is, decarbonize. The rates of conversion to sinks cannot be controlled.

Addressing the issue of low-carbon growth is a challenge for all countries, but particularly for developing countries since low carbon is often associated with low growth. However, this is not necessarily so and developing countries need to find ways to grow with low-carbon intensity. In the context of industry, higher energy efficiency is the way to achieving low-carbon growth. This translates to specific processes in specific industries, and benchmarking industry to international energy-efficient standards and processes. SMEs also need to be included in this exercise. The regulatory structure also needs to be geared towards such practices and monitoring and implementation of best practices need to be undertaken. Given the interdependencies between various sectors of the economy, the potential for increasing energy efficiency needs to be identified and implemented across all sectors, in households, transport as well as cities.

For a country like India, the issue of energy efficiency in industries has been adequately addressed in most of the energy-intensive industrial sectors, and the regulatory mechanism to make this transition is also in place. However, the main problem arises in generation of power, in which India is still largely dependent on coal. This is the main challenge that Indian policy makers face and all efforts at decarbonizing the economy have to tackle this problem head on. Efforts at increasing the use of renewable energy, hence need to be implemented on a war footing and India needs support of the developed countries in deploying some of the latest technology in this arena in the country.

References

Ajmi AN., Hammoudeh S, Nguyen DC, Sato JR (2015) On the relationship between CO_2 emissions, energy consumption and income: the importance of time variation. Energy Economics. (forthcoming)

Banerjee R, Cong Y, Gielen D, Jannnuzzi G, Marechal F, McKane AT, Rosen MA, van Es D, Worrell E (2012) Chapter 8—energy end-use: industry. In: Cambridge UK, New York NY (eds) Global energy assessment-toward a sustainable future. Cambridge University Press, International Institute for applied Systems Analysis, USA, Laxenburg, Austria, pp 513–574

CII (2013) Widening the coverage of the PAT scheme: Indian chemical industry, CII, December

Ghosh S (2010) Examining carbon emissions economic growth nexus for India: a multivariate cointegration approach. Energy Policy 38:3008–3014

Grubler A, Bai X, Buettner T, Dhakal S, Fisk D, Ichinose T, Keirstead J, Sammer G, Satterthwaite D, Schultz N, Shah N, Steinberger J, Weisz H (2012) Urban energy systems. in global energy assessment: toward a sutainable future. Cambridge University Press, Cambridge, UK, New York, NY, USA, and the International Institute for Applied Systems Analysis, Laxenburg, Austria, pp 1307–1400

Guttikunda SK, Jawahar P (2012) Atmos Environ 62:551–561

Holian MJ (2014) The effect of social and economic development on air pollution in Indian Cities. Environ Urban Asia 5(1):1–15

International Energy Agency (2007) Tracking industrial energy efficiency and CO_2 emissions. IEA/ OECD, Paris, France

International Energy Agency (2008) World energy outlook. OECD/IEA, Paris, France

International Energy Agency (2013) Redrawing the energy-climate map. IEA/OECD, Paris, France

International Energy Agency (2014) World energy investment outlook. IEA/ OECD, Paris, France

International Energy Agency (2015) World energy trends: excerpt from energy balances of non-OECD countries, IEA

O'Neill C, Ren X, Jiang L, Dalton M (2012) The effect of urbanization on energy use in India and China in the iPETS model. Energy Economics 34:S339–S345

Ramachandra TV, Aithel BH, Sreejith K (2015) GHG footprint of major Cities in India. Renew Sustain Energy Rev 44:473–495

Ray S, Goldar A, Saluja S (2014) Feedstock for the petrochemical industry, ICRIER Working Paper 271

Reddy MS, Venkataraman C (2002) Atmos Environ 36:677–697

Sathaye J, Prioce L, Rue du Can S, Fridley F (2005) Assessment of energy use and energy savings potential in selected industrial sectors in India. Lawrence Berkeley National Laboratory (LBNL-57293), Berkeley, CA,

Seto KC, Guneralp B, Hutyra LR (2012) Global Forecasts of Urban Expansion to 2030 and direct impacts on biodiversity and carbon pools. Proc Natl Acad Sci USA 109(40):16083–16088

Seto KC, Dhakal S, Blanco H, Dewar D, Huang L, Inaba A, Kansal A, Lwasa S, McMahon JE, Muller DB, Murakami J, Nagendra H, Ramaswami A (2014) Human settlements, infrastructure, and spatial planning. In Edenhofer O, Pichs-Madruga R, Sokona Y, Farahani E, Kadner S, Seyboth K, Adler A, Baum I, Brunner S, Eickemeier P, Kriemann B, Savolainene J, Schlomer S, von Stechow C, Zwicke T, Minx JC (eds.), Climate change 2014: mitigation of climate change. Contribution of working group III to the fifth assessment report of the intergovernmental panel on climate Change. Cambridge University Press, Cambridge, United Kingdom, New York, USA

UNEP (2013) The Emissions Gap Report 2013—A UNEP Synthesis Report. Nairobi, Kenya, UNEP

UNEP (2014) The Emissions Gap Report 2014 – A UNEP Synthesis Report. Nairobi, Kenya, UNEP

UNIDO (2010) Global Industrial Energy Efficiency Benchmarking: An Energy Policy Tool, Working paper

US Energy Information Administration (2013) International Energy Outlook 2013

US Energy Information Administration (2014) International Energy Outlook 2014

World Bank (2012) Turn down the Heat: Why a 4 ° C warmer world must be avoided International Bank for Reconstruction and Development. The World Bank, Washington DC

Chapter 8
Decarbonization of Indian Railways

T. S. Ramakrishnan

Abstract About 21% of worldwide CO_2 emissions are attributed to transport in 2014. India's rail transport has the lowest rate of energy consumption per Tonne-Kilometre (TKM) for goods by rail (102 kJ/TKM) among the countries considered (OECD/IEA 2015). However, given the increase in demand owing to the growth in population and GDP as also the Indian Railways' expansion plans, the energy consumption of Indian Railways is set to rise. Indian Railways has already considered the importance of increasing the share of low-carbon renewable energy sources such as solar and wind in its total energy mix. It is opportune for the Indian Railways to cut CO_2 emissions using electricity generated by the sun, wind or biofuels. This chapter is an outcome of a study which examined the feasibility of complete decarbonization of Indian Railways under different scenarios for achieving this goal. The study estimated first, rail passenger (both suburban and non-suburban transport) and freight forecast between 2015–16 and 2030–31. As the second step, the study estimated the expected share of electrified tracks in the overall running tracks and the expected share of electric hauling of passenger and freight transport, between 2015–16 and 2030–31. In the third step, this study calculated the energy forecast in terms of electricity and diesel arising from rail transport forecast and share of electric and diesel traction was estimated, between 2015–16 and 2030–31. As the fourth step, the CO_2 emissions attributed to Indian Railways were also calculated, between 2015–16 and 2030–31.

8.1 Introduction

India's development challenges are manifold. It faces a burgeoning population, and is expected to become the most populous country in the world, overtaking China in 2025. Concomitant with its population growth is its energy consumption, making it the fifth largest greenhouse gas (GHG) emitter in the world (MoEF 2010). While

T. S. Ramakrishnan (✉)
Consultant and Researcher, Bangalore, India
e-mail: ramakrishnan@iima.ac.in

R. Kathuria et al. (eds.), *Low Carbon Pathways for Growth in India*, India Studies in Business and Economics, https://doi.org/10.1007/978-981-13-0905-2_8

India needs to overcome the socioeconomic challenge of poverty, infrastructure cannot be provided at the cost of its environment. The challenge that India faces is unique: it should grow fast, but not in an unsustainable manner.

About 21% of worldwide CO_2 emissions are attributed to transport in 2014 (World Bank n.d). The carbon intensity of travel depends not only on technical issues such as characteristics of vehicles and the fuels used but also the travel mode choices made by people, given the availability of travel modes for them. Though cars and trucks represent the bulk of carbon emissions from transport, the railways also are critical. There is a silver lining in the rail transport in India that it has the lowest rate of energy consumption per Tonne-Kilometre (TKM) for goods by rail (102 kJ/TKM) among the countries considered (OECD/IEA 2015). However, given the increase in demand owing to the growth in population and GDP as also the Indian Railways' expansion plans, the energy consumption of Indian Railways is set to rise. In FY 2011–12, Indian railways catered to 1047 billion Passenger Kilo Metre (PKM) of passenger transport, which constituted about 10% of total passenger transport in India (NTDPC 2014). In FY 2011–12, Indian railways catered to 668 revenue earning billion TKM, which constituted about 33% of total freight transport in India (NTDPC 2014). India Transport Report—Moving India to 2032 set an ambitious target of 6559 billion TKM of freight transport for railways in 2031–32, which will constitute 50% modal share for rail in freight transport and 5765 Billion PKM of passenger transport for railways in 2031–32 (NTDPC 2014). So, this is a challenge and an opportunity for Indian Railways as India could increase its freight transport and its modal share in freight transport and thereby leverage the low-carbon footprint associated with rail transport by sourcing the energy for its operations from renewable energy sources.

The entire freight movement by truck depends on diesel, whereas about 36% of rail freight transport in 2011–12 depended on diesel, which is an air pollutant. Indian Railways hauled two-thirds of its freight traffic and half of its passenger traffic by electricity although 42% of its network length (42,000 km) is electrified in 2015–16 (PIB November 3, 2016), which corresponds to about 46% of the running track kilometre. IR ambitiously targeted electrification of 2000, 4000, 6000, 6200 and 6200 km of running track in 2016–17, 2017–18, 2018–19, 2019–20 and 2020–21 respectively (PIB November 3 2016).

In this backdrop, Indian Railways has already considered the importance of increasing the share of low-carbon renewable energy sources such as solar and wind in its total energy mix. Indian Railways consumed about 18 billion units of electricity in 2015–16 and its electricity requirement will increase to 49 billion units by 2030 (PIB November 3 2016). Indian government made a commitment to reduce its greenhouse gas emissions by 33–35% by 2030 from the 2005 levels in Conference of Parties held in Paris. Decarbonization of Indian Railways will help achieve this commitment made by India. In that process, Indian Railways has an internal target of installing 1000 MW of solar power, out of which 500 MW on rooftops and balance 500 MW on land and 500 MW of wind power by 2020. It also set a target of 40% of its cumulative electricity generation capacity through renewable energy by 2030 (PIB November 3 2016). However, even if the renewable energy target is achieved, the remaining electric energy will have to come from general pool of electricity. In 2012, coal-fired electricity accounted for 60% of India's installed capacity and 71% of its electricity generation (Office of the Chief Economist 2015) and hence, electrification and decarbonization of railways should consider the investment already made into nonrenewable power generation plants.

However, it is opportune for the Indian Railways to cut CO_2 emissions by using electricity generated by the sun, wind or biofuels. The Indian government has announced that an installed renewable capacity of 175 GW by 2022 is targeted. This includes 100 GW of solar energy and 23 GW of wind energy. The targets of the National Solar Mission can be achieved if the complex set of issues related to land acquisition, network expansion, remuneration and financing is overcome. The biggest challenge of land acquisition has been solved by the Railways since it has identified land parcels around unused tracks where the solar mission can be implemented.[1] The Solar Mission aims to generate 60 GW of the 100 GW target by utility-scale plants, while the remainder, 40 GW, will come from rooftop Photo Voltaic (PV) installations and other small-scale and off-grid installations. Here too, the Indian Railways has an advantage and is in the process of identifying buildings suitable for rooftop installations.

This chapter is an outcome of a study which examined the feasibility of complete decarbonization of Indian Railways under different scenarios for achieving this goal. The study estimated, first, rail passenger (both suburban and non-suburban transport) and freight forecast between 2015–16 and 2030–31 by considering the available source of information that gives indication of supply and demand of rail transport. As the second step, the expected share of electrified tracks in the overall running tracks and the expected share of electric hauling of passenger and freight transport, both based on available source of information, between 2015–16 and 2030–31, were estimated. As a third step, the energy forecast in terms of electricity and diesel arising from rail transport forecast and share of electric and diesel traction was estimated, between 2015–16 and 2030–31. As the fourth step, the CO_2 emissions attributed to Indian Railways were also calculated, between 2015–16 and 2030–31.

[1]The Railways Minister, Shri Suresh Prabhu announced in his budget speech on 26 February 2015 'To reduce dependence on fossil fuels, it is intended to expand sourcing of Solar Power as part of the Solar Mission of Railways. Work is in full swing at the solar power plant at Katra station and is slated for completion in March 2015. Further, 1000 MW solar plants will be set up by the developers on Railway/private land and on rooftop of Railway buildings at their own cost with subsidy/viability gap funding support of Ministry of Non-Renewable Energy in next five years'.

Table 8.1 Rail Passenger transport demand as forecast by NTDPC up to 2031–32

Year	Rail Passenger Traffic (BPKM)
2011–12	1047
2016–17	1509
2021–22	2300
2026–27	3596
2031–32	5765
Rail passenger transport CAGR (%) between 2011–12 and 2031–32	8.91
GDP CAGR (%) between 2011–12 and 2031–32	8.10
Elasticity of rail passenger transport	1.10

Source NTDPC (2014)

8.2 Passenger and Freight Transport Forecast of Indian Railways up to 2031–32

To estimate the rail passenger and freight transport forecast, it is imperative to look at previous such estimates, if any. The NTDPC[2] estimated Indian Railways' passenger and freight forecast between 2011–12 and 2031–32. NTDPC took 2011–12 as the base year and estimated the rail passenger and freight transport forecast up to 2031–32 in a slab of 5 years each. The Compounded Annual Growth Rates (CAGR) of GDP assumed by the NTDPC in their estimates is as follows (NTDPC 2014).

- 6.9% between 2011–12 and 2015–16
- 8% between 2016–17 and 2020–21
- 8.5% between 2021–22 and 2025–26
- 9% between 2026–27 and 2031–32

For the period between 2011–12 and 2031–32, NTDPC assumed that the rail passengers demand elasticity of 1.10 for every 1% CAGR of GDP and accordingly, NTDPC estimated that the rail passenger transport in 2031–32 would be 5765 BPKM as shown in Table 8.1.

NTDPC assumed an elasticity of 1.49 for rail freight transport demand between 2011–12 and 2031–32 for every 1% CAGR of GDP and accordingly, NTDPC estimated that the rail freight transport in 2031–32 would be 6559 BTKM as shown in Table 8.2.

However, the NTDPC estimates did not consider the supply side constraints of catering to additional rail passenger and freight transport demand and was more a

[2]'India Transport Report—Moving India to 2032' submitted by the National Transport Development Policy Committee (NTDPC) and headed by Rakesh Mohan in 2014 estimated passenger and freight demand.

Table 8.2 Rail passenger and freight transport demand as forecast by NTDPC up to 2031–32

Year	Revenue earning rail freight traffic (BTKM)
2011–12	668
2016–17	1070
2021–22	1885
2026–27	3535
2031–32	6559
Rail freight transport CAGR (%) between 2011–12 and 2031–32	12.10
GDP CAGR (%) between 2011–12 and 2031–32	8.10
Elasticity of rail freight transport	1.49

Source NTDPC (2014)

target for Indian Railways. Moreover, the actual passenger and freight traffic carried out Indian Railways up to 2014–15 is available and hence the estimates need to be done only between 2015–16 and 2031–32. Accordingly, here, the rail passenger and freight transport forecast was estimated between 2015–16 and 2030–31 for three different scenarios.

- Optimistic scenario—CAGR of 9% between 2015–16 and 2031–32
- Realistic scenario—CAGR of 8% between 2015–16 and 2031–32
- Pessimistic scenario—CAGR of 7% between 2015–16 and 2031–32

8.2.1 *Passenger and Freight Transport Forecast of Indian Railways up to 2031–32*

It is found that the average elasticity of passenger transport across all modes in India between 1950–51 and 2010–11 was 1.50 and as per World Bank estimates, developing countries will have a freight transport elasticity of 1.25. Indian Railways has been suffering from capacity constraints and which formed the primary reason for its inability to augment passenger and freight transport. Between 1950–51 and 2014–15, while the freight loading has grown by 1344% and passenger kilometres by 1642%, the Route kilometres have grown by only 23% and Doubling & Multiple route length by only 289% (MoR 2015). As a result, it does not have the capacity to cater to the demand of passenger and freight transport and hence the rail passenger and freight elasticities are very low in comparison with the overall passenger and freight transport elasticities. The second reason is that whatever additional track infrastructure that IR added was not in the High-Density Corridors. By 2009, more than 55% of the traffic had moved along the railway golden quadrilateral and its diagonals, connecting Delhi, Kolkata, Mumbai and Chennai

Table 8.3 Line capacity status of overall network and high-density network on IR

Load (%)	<80	80–100	100–120	120–150	>150	Total
Overall IR network sections	415	228	193	66	84	1219
High-density network sections	30	56	59	79	23	247

Source MoR (2015)

(67% of them have surpassed 100% capacity utilization), and the Delhi-Guwahati route, which formed less than 20% of the total IR network (MoR 2009). The heavy congestion in high-density corridors has not allowed IR to cater the passenger and freight travel demand made the sections of High-Density Corridors to carry rail traffic beyond its capacity as shown in Table 8.3.

The third reason is that IR did not overhaul its operations based on the changing passenger and freight demand and has been catering to long distance, medium distance and short distance passengers all at once and as a result run trains with huge speed differentials. As a result, IR has been unable to improve its throughput and whatever increase in throughput comes at the cost of overloading its network.

The rail passenger and freight elasticity between 2010–11 and 2014–15 was 0.58 and 0.32, respectively, and this was essentially due to supply-side constraints. Although IR has stepped up its construction of additional infrastructure since 2014–15, it is expected that total effective running track available for passenger transport and freight transport to 95,803 km in 2019–20 from 90,803 km in 2014–15. If the Business As Usual (BAU) scenario is assumed, which means the increase in carrying capacity of IR will be increasing in direct proportion to increase in running track length between 2014–15 and 2019–20, IR will have the capacity to achieve only an elasticity of 0.14 in passenger transport between 2014–15 and 2019–20 for realistic scenario of GDP growth. However, either by further loading the already overloaded tracks or bringing some improvement in scheduling of trains and operational efficiency, it is forecast that IR may be able to achieve 1506 billion PKM in 2019–20 from 1147 billion PKM in 2014–15 for realistic scenario of 8% GDP CAGR. This would put the CAGR of rail passenger growth between 2014–15 and 2019–20 at 5.60% with the corresponding elasticity of 0.7 with respect to GDP CAGR. In case of freight traffic, on improving efficiency of network operations and other measures, it is forecast that IR may be able to achieve 830 billion TKM in 2019–20 from 682 billion TKM in 2014–15 in revenue earning freight traffic for realistic scenario of 8% GDP CAGR. This would put the CAGR of rail freight growth between 2014–15 and 2019–20 at 4.01% with the corresponding elasticity of 0.5 with respect to GDP CAGR.

However, the period between 2020–21 and 2024–25 would be remarkably different in terms of improvement in supply side. IR has been progressing on Western Dedicated Freight Corridor (WDFC) and Eastern Dedicated Freight Corridor (EDFC) and both these DFC are expected to be commissioned by December 2019. On the one hand, these corridors would cater to huge additional freight traffic, as freight trains in DFC are expected to run at a maximum speed of

100 kmph and an average speed of 75 kmph and hence relieve 33% of the freight traffic in the existing conventional lines to passenger traffic. National High Speed Rail Corporation Limited is already constituted for the construction of HSR between Ahmedabad and Mumbai and the first HSR of India is expected to be commissioned by 2023. Given this, it is expected that the total effective running track available for passenger transport would be 108,561 km in 2024–25 from 95,803 km in 2019–20 and the total effective running track available for freight transport would be 121,329 km in 2024–25 from 95,803 km in 2019–20. In addition to increase in total effective running track, the substantial improvement in the operations of conventional rail network also would increase the capacity of IR to cater to higher passenger and freight transport and in this backdrop, it is forecast that IR may be able to achieve 2561 billion PKM in 2024–25 from 1506 billion PKM in passenger transport in 2019–20 for realistic scenario of 8% GDP CAGR. This would put the CAGR of rail passenger growth between 2019–20 and 2024–25 at 11.20% with the corresponding elasticity of 1.4 with respect to GDP CAGR. It is also forecast that IR may be able to achieve 1997 billion TKM in 2024–25 from 830 billion TKM in 2019–20 in revenue earning freight transport for realistic scenario of 8% GDP CAGR. This would put the CAGR of rail freight growth between 2019–20 and 2024–25 at 19.2% with the corresponding elasticity of 2.4 with respect to GDP CAGR.

The feasibility study for the construction of additional Dedicated Freight Corridors has been initiated for the following corridors (PIB Aug 7 2015):

- East–West Corridor Kolkata Mumbai 2330 km
- North–South Corridor Delhi Chennai 2343 km
- East Coast Corridor Kharagpur Vijayawada 1110 km
- Southern Corridor Chennai Goa 899 km

Although the feasibility study was initiated, the above DFC projects may take about 10–14 years for completion and hence it is assumed here that these DFC corridors would be commissioned between 2025–26 and 2030–31.

Although, IR planned to develop HSR in the six routes amounting to 3500 running track km to begin with, it now conducted feasibility study for Golden Quadrilateral and its diagonals except Chennai–Kolkata route. There are no time-lines set for these HSR projects as the execution of these projects depends on various ambiguous factors. Given that India's first HSR line may be commissioned in 2023, it is assumed here that 2000 route km and 4000 running track km of HSR may be available for passenger traffic by 2030–31. In addition to full-fledged HSR, IR also identified certain routes such as Chennai–Bangalore–Mysore, Delhi–Chandigarh, New Delhi–Howrah and New Delhi–Mumbai for semi-HSR corridors, where the conventional rail tracks would be upgraded to run trains at 160–200 kmph.

Given this, it is expected that the total effective running track available for passenger transport would be 134,415 km in 2030–31 from 108,561 km in 2024–25 and the total effective running track available for freight transport would be

176,007 km in 2030–31 from 121,329 km in 2024–25. Given this, IR will have the capacity to achieve an elasticity of 1.60 in passenger transport and 2.50 in freight transport between 2024–25 and 2030–31 for realistic scenario of GDP growth. With the huge addition of HSR and DFC as well as major hauling by electric traction both for passenger and freight transport, it is forecast here that IR may be able to achieve 5276 billion PKM in 2030–31 from 2561 billion PKM in passenger transport in 2024–25 for realistic scenario of 8% GDP CAGR. This would put the CAGR of rail passenger transport growth between 2024–25 and 2030–31 at 12.8% with the corresponding elasticity of 1.1 with respect to GDP CAGR. It is also forecast that IR may be able to achieve 5276 billion TKM in 2030–31 from 2561 billion TKM in 2024–25 in revenue earning freight transport for realistic scenario of 8% GDP CAGR. This would put the CAGR of rail freight growth between 2024–25 and 2030–31 at 20.0% with the corresponding elasticity of 2.5 with respect to GDP CAGR.

The NTDPC report took uniform value of elasticity for both rail passenger and freight transport, and forecasted transport growth accordingly without considering supply-side constraints. However, here, the rail transport growth was forecast considering the supply side aspects very much. However, the overall growth forecast for both rail passenger and freight transport between 2011–12 and 2030–31 was kept almost equal to what NTDPC prescribed.

Based on the above discussion, the rail passenger and freight transport forecast for three different scenarios up to 2031–32 is shown in Tables 8.4 and 8.5, respectively.

Table 8.4 Rail passenger transport forecast estimated for three different scenarios up to 2031–32

Year	Passenger transport in BPKM	Remarks
2010–11	979	Actual
2011–12	1047	
2012–13	1098	
2013–14	1140	
2014–15	1147	
CAGR (%) of rail passenger transport between 2010–11 and 2014–15	4.04	
CAGR GDP between 2010–11 and 2014–15	7	
Actual elasticity of rail transport between 2010–11 and 2014–15	0.58	

(continued)

Table 8.4 (continued)

Year	Passenger transport in BPKM			Remarks
	Optimistic scenario	Realistic scenario	Pessimistic scenario	Remarks
CAGR of GDP (%) between 2015–16 and 2031–32	9	8	7	Assumed
Elasticity of rail passenger transport between 2014–15 and 2019–20	0.7	0.7	0.7	
Elasticity of rail passenger transport between 2020–21 and 2024–25	1.4	1.4	1.4	
Elasticity of rail passenger transport between 2025–26 and 2031–32	1.6	1.6	1.6	
	Passenger transport in BPKM			Remarks
2019–20[a]	1557	1506	1457	Forecast
2024–25	2818	2561	2325	
2030–31	6317	5276	4396	
CAGR (%) of rail passenger transport between 2010–11 and 2030–31	9.77	8.79	7.8	
Effective elasticity of rail passenger transport between 2010–11 and 2030–31	1.1	1.1	1.1	
Additional rail passenger transport between 2014–15 and 2031–32	5170	5276	4396	

Source Author's calculations based on MoR (n.d.a), MoR (n.d.b) and Statistics Times (Feb 18 2016)

[a]Although the passenger traffic was estimated for the years from 2010–11 to 2030–31, the actual data were available for passenger traffic between 2010–11 and 2014–15. The passenger transport in 2014–15 was 1147 BPKM. For the optimistic scenario of 9% GDP CAGR and elasticity of 0.7, the annual growth of passenger traffic will be 9*0.7 = 6.3 and by 2019–20, the passenger traffic would be 1147*(1.063)^5 = 1557 BPKM

Table 8.5 Estimation of rail freight forecast for three different scenarios up to 2031–32

Year	Revenue earning freight transport in BTKM	Remarks
2010–11	625	Actual
2011–12	668	
2012–13	650	
2013–14	666	
2014–15	682	
CAGR of rail freight transport between 2010–11 and 2014–15 (%)	2.21	
CAGR GDP between 2010–11 and 2014–15 (%)	7	
Elasticity of rail transport between 2010–11 and 2014–15	0.32	

(continued)

Table 8.5 (continued)

Year	Revenue earning freight transport in BTKM			Remarks
	Optimistic scenario	Realistic scenario	Pessimistic scenario	Remarks
CAGR of GDP (%) between 2015–16 and 2031–32	9	8	7	Assumed
Elasticity of rail freight transport between 2010–11 and 2019–20	0.5	0.5	0.5	
Elasticity of rail freight transport between 2020–21 and 2024–25	2.4	2.4	2.4	
Elasticity of rail freight transport between 2025–26 and 2031–32	2.5	2.5	2.5	
	Revenue earning freight transport in BPKM			Remarks
2019–20[a]	850	830	810	Forecast
2024–25	2260	1997	1761	
2030–31	7637	5963	4634	
CAGR (%) between 2010–11 and 2030–31	13.33	11.94	10.54	
Effective elasticity of rail freight transport between 2010–11 and 2031–32	1.48	1.49	1.51	
Additional rail freight transport between 2014–15 and 2031–32	6955	5281	3952	

Source Author's calculations based on MoR (n.d.a), MoR (n.d.b) and Statistics Times (Feb 18 2016)

[a]Although the freight traffic was estimated for the years from 2010–11 to 2030–31, the actual data were available for freight traffic between 2010–11 and 2014–15. The revenue earning freight transport in 2014–15 was 682 BTKM. For the optimistic scenario of 9% GDP CAGR and elasticity of 0.5, the annual growth of passenger traffic will be 9*0.5 = 4.5 and by 2019–20, the freight traffic would be 682*(1.045)^5 = 850 BTKM

8.3 Estimates of Energy Requirement of Indian Railways

8.3.1 IR's Total Energy Requirement

IR's current and future total energy requirement has two major components.

(i) Energy for non-transport activities
(ii) Energy for transport activities

The energy constituted INR 29,293 crores, which was about 19% of the total expenditure of IR in 2013–14. About 50% of IR's passenger transport and about two-thirds of its freight transport have been hauled by electricity and non-transport energy needs are essentially met by electricity and the share of electricity in the total fuel consumption is increasing.

8.3.1.1 Energy for Non-Transport Activities

The energy for non-transport is essentially refers to electricity consumed for all activities other than traction. Although, nonelectric energy may be used in some cases for non-transport activities, energy requirement for non-transport activities are essentially catered by electricity and hence the discussion of non-transport energy confined to electrical energy here. Energy for non-transport includes consumption of electricity in the manufacturing units, railway offices, railway stations and locomotive sheds. IR consumed 2411 Million Units (MU) of electricity for non-transport purposes in 2007–08, which formed a share of 17.1% in the total electricity consumption of IR. It increased 2503 MU in 2014–15, which formed a share of 13.72% in the total electricity consumption of IR. The absolute electrical energy for non-transport in 2014–15 was 92 MU higher than that of 2007–08. The aggressive energy efficiency measures implemented by IR such as energy audits, adoption of LED lighting, super-efficient ceiling fans for railway quarters, offices, railway stations, Energy Management System such as ISO: 9001 and ISO: 14,000, use of energy efficient water coolers, pump automation with GSM-based techniques, use of energy efficient pumps, microcontroller-based Automatic Platform Lighting Management System with segregation of 70/30% circuits, use of 3 stars and above-labelled electrical products and equipment, solar-based LED lighting system for level crossing gates, use of solar water heater in place of electric geyser, occupancy sensors in offices, 14 Lakhs CFL given free of cost by IR to its employees in the year 2009–10, adoption of Energy Conservation Building Code (ECBC) 2007 for new buildings (Indian Railway Institute of Civil Engineers—IRICEN New Administrative Green Building was constructed with ECBC norms, which saved 46% of energy) contributed to the substantial reduction of non-transport electrical energy consumption.

8.3.1.2 Energy for Transport Activities

The energy for rail transport is composed of:

1. Energy for passenger transport
2. Energy for freight transport

The energy consumption of either passenger or freight trains depends on two factors, the energy consumed for compensating the air drag, and the energy consumed in accelerating to the cruising speed. Heavier and faster trains need more energy to accelerate. That is when the ratio of payload to the total weight (total weight is the sum of payload and tare weight) is lower, the energy required for one unit of passenger transport or freight transport would be higher.

However, the energy for rail transport may also be classified into electrical energy and diesel energy, depending on whether the hauling is done by electric traction or diesel traction. The energy for freight transport is essentially for hauling freight trains, whereas the energy for passenger transport consists of the in-vehicle consumption of energy for hauling as well as for in-vehicle consumption of passenger carriages.

8.3.2 *The Key Aspects of Energy Consumption of Indian Railways*

8.3.2.1 Electric Locomotives Versus Diesel Locomotives

Indian Railways uses both electric and diesel locomotives for non-suburban passenger and freight transport and Electric Multiple Units (EMU) and Diesel Multiple Units (DMU)/Diesel Hydraulic Multiple Units (DHMU) for suburban passenger transport. IR has been increasing its electric locomotives consistently over the years. In 1950–51, Indian Railways had 8120 steam locomotives, 17 diesel and 72 electric locomotives.[3] In 2013–14, this figure was 43 for steam, 5633 for diesel and 4823 for electric locomotives. In 2014–15, Indian Railways had a fleet of 10,749 locomotives, of which 5749 are diesel and 5000 are electric. Among the electric locomotives, 16% are high horsepower (HHP), 3-phase locomotives (MoR 2015).

Electric locomotives are lighter and have fewer moving parts compared to diesel ones, though diesel locomotives have higher tractive power and the ability to start heavier trains. Indian Railways uses modern diesel locomotives based on an integrated gate bipolar transistor (IGBT).[4] For the same weight, a diesel locomotive can haul a 10% higher load compared to electric locomotives. The major advantage of diesel locomotives is that they can work seamlessly under all conditions and terrains. The main limitation is that diesel locomotives should be fueled at regular intervals and elaborate arrangements have to be made for moving and storing fuel. By contrast, electric locomotives can run on electricity from any source (coal, gas, diesel, renewable, etc.). Electric locomotives can run only on electrified sections of a railway network. The diesel locomotive is the primary producer of power and as diesel burns, it is converted to electric energy for traction purposes. In the case of electric locomotives, there is no visible pollution except at the site of electricity generation or the power plant.[5] Gangwar and Sharma (2014) note that while the efficiency of diesel engines is about 35%, for electric traction (considering a grid-to-loco efficiency of 98%, power plant efficiency of 30–35%, 10% transmission losses and 2% losses in locomotive), it is about 28–30%. Given the current

[3]Ministry of Railways (Railway Board) (2015a).

[4]200 such locomotives of 9000 HP are being procured for maintenance of infrastructure at Rewari on the Western Dedicated Freight Corridor.

[5]Kathpal (1997) notes that the major pollutants from diesel locomotives are NO_x and particulate matter, while thermal power plants produce suspended particulate matter, oxides of nitrogen (NO_x), sulphur dioxide and ash.

generation and distribution inefficiency of electricity, the end-to end energy efficiency of diesel locomotives is somewhat better than that of electric locomotives. However, with substantial improvement in generation and distribution of electricity, the efficiency of electric locomotives is bound to increase substantially.

8.3.2.2 Electrical Energy Procurement and Usage for Indian Railways

While the Indian Railways has procured electricity in bulk from different states, the state-owned distribution companies have been charging Indian Railways at industrial rates. According to the Central Electricity Regulatory Authority (CERC), the Indian Railways has been given the status of 'deemed distribution licensee'[6] allowing it to procure power relatively cheaply, and avoid the cross-subsidy surcharge. This would translate to annualized savings of Rs. 1300 crore (Railway Budget 2016–17).[7]

Of the 65, 808 route kilometres in 2013–14, 33%, or 21, 614 route kilometres are electrified[8] 26, 269 RKM of Indian Railways tracks (39.9%) are electrified and 65.4% of freight traffic and 51.2% of passenger traffic are hauled on electric traction: this makes up 36.3% of the total cost of traction fuel,[9] which means IR saves expenditure cost when it uses electric traction for passenger and freight transport. However, Gangwar and Sharma (2014) reported that the power sector generates about 38% of carbon emissions as most of the generating plants are coal or gas based. Fuel usage in transport resulted in 142 MT of CO_2 equivalent in 2007, while the railways released 7.12 MT of CO_2 eq. of GHGs (Parikh et al. 2014). The Indian Railways depends heavily on energy obtained by burning fossil fuels—either diesel, or electricity produced in coal-based power plants.

8.3.2.3 Electrical Energy Consumption of Indian Railways

The Indian Railways are the largest single organization with an electricity consumption share of 2.4% of total electricity generated in the country. In 2007–08, they consumed 14.1 billion kilowatt-hours (kWh), of which 11.7 billion kWh (83%) was for traction, while 2.4 billion kWh was for non-traction purposes (UNDP 2015). The Indian Railways consumed 17.5 billion units in 2013–14 (1.8% of the country's power generated), of which 15 billion units were for traction and

[6]http://www.financialexpress.com/article/budget-2016/rail-budget-2016-indian-railways-saves-big-on-fuel-cost-especially-electricity/216262/.

[7]http://www.indianrailways.gov.in/railwayboard/uploads/directorate/finance_budget/Budget_2016-17/RailBudgetSpeech_2016-17_Eng.pdf.

[8]According to the Outcome and Performance Budget of railways for 2015–16 report of the Ministry of Railways, 24, 891 route kilometre (RKM) were electrified in the total railway network of 65, 436 RKM. The map in the appendix shows the electrified sections in red.

[9]http://pib.nic.in/newsite/PrintRelease.aspx?relid=122515.

2.5 billion units were for non-traction.[10] The total annual bill for electricity was Rs. 11,300 crores, of which 9650 crores were for traction and 1650 crores were for non-traction. By March 2014, 24,891 Route Kilometre (cumulative) were electrified, which constituted 38% of the total rail network and 43.5% of the Broad Gauge (BG) system.

The IR has initiated an ambitious plan for electrification; 36% of freight and 53% of passenger traffic currently hauled by diesel locomotives will switch over to electricity. In 2014–15, 1375 RKM were electrified, which is the highest ever in a year.[11] It is envisaged that 80% of rail freight and 60% of passenger traffic will run on electrical energy by 2031–32.

8.3.3 IR's Total Electrical Energy Requirement for Transport

Energy consumption for train operations is measured in per tonne-km, train-km, car-km, seat-km or per passenger-km (Andersson and Lukasezewicz 2006). In case of the last, energy consumption is related to the number of passengers using the actual train or the number of passenger-km produced. However, electrical energy consumption data for rail transport in terms of train-km or car-km or seat-km or PKM of various classes or TKM are not available for Indian Railways. The only data available are based on the Gross Tonne Kilometre (GTKM) moved. GTKM here refers to the product of the total mass of the train moved and the distance travelled. In freight trains, the in-vehicle consumption is almost nil so all the electrical energy drawn is used to move the train. However, in case of passenger trains, this includes in-vehicle electricity consumption of passenger trains as the energy required for operating lights, fans and air conditioners essentially comes from the traction energy. In addition, the ratio of payload to tare weight is not equal across suburban and non-suburban trains. Among the non-suburban trains, the ratio of payload to tare weight is lower for the AC classes (such as 1AC, 2AC, 3AC, Executive Class and Chair Car, in that order) over the non-AC classes (sleeper class, seating, unreserved), and hence the electrical energy required to haul one PKM of AC class is higher than that required to move non-AC classes. It is obvious that air-conditioned coaches consume more energy than the fans of the non-AC coaches.

The total electric energy consumed by IR at present for traction is measured as per Eq. (8.1).

[10]Ministry of Railways (Railway Board) (2015b).

[11]http://pib.nic.in/newsite/PrintRelease.aspx?relid=122515.

$$\text{Total electric energy consumed by IR for traction} = \text{Electric energy consumed by IR for freight transport} + \text{Electric energy consumed by IR for passenger transport} \quad (8.1)$$

where Electric energy consumed by IR for passenger transport = Electric energy consumed by IR for in-vehicle use of passenger transport + Electric energy consumed by IR for hauling of passenger transport.

where Electric energy consumed by IR for in-vehicle use of passenger transport = Electric energy consumed by IR for in-vehicle use of non-suburban AC coaches + Electric energy consumed by IR for in-vehicle use of non-suburban non-AC coaches + Electric energy consumed by IR for in-vehicle use of suburban non-AC coaches.

where Electric energy consumed by IR for hauling of passenger transport = Electric energy consumed by IR for hauling of non-suburban AC coaches + Electric energy consumed by IR for hauling of non-suburban non-AC coaches + Electric energy consumed by IR for hauling of suburban non-AC coaches.

IR planned to introduce AC coaches for suburban passenger transport. In that case, one more component of electric energy consumed by IR for in-vehicle use of suburban AC coaches would be added to the Electric energy consumed by IR for in-vehicle use of passenger transport.

8.3.4 *Dissecting the Energy Requirement for Passenger and Freight Transport*

As there is no direct data available for estimating the energy consumed for one unit of passenger transport in suburban and non-suburban trains as well as in AC and non-AC classes or for one unit of freight transport, a four-step method was devised to calculate to estimate the energy required for the same. This involved estimation of in-vehicle electricity consumption, electricity energy consumed for hauling and electrical energy consumed in the hauling of non-suburban and suburban passenger transport. Estimation of future electrical traction in rail passenger and freight transport also included the expected use of electric traction in hauling non-suburban passenger and freight transport. Based on these calculations, the electricity requirement of the Indian Railways has been projected till 2030–31.

8.3.4.1 First Step: Estimation of in-Vehicle Electricity Consumption Per PKM of Passenger Transport

The in-vehicle electricity consumption per PKM of passenger transport varies depending on the nature of electricity use and the average number of passengers accommodated in a passenger carriage. There are totally ten classes in IR with different accommodation pattern and varied in-vehicle electricity energy use. However, the major difference in in-vehicle electricity consumption emanates from whether the coach is air-conditioned or not. Hence, here, the non-suburban transport has been classified as AC and non-AC classes and no further split is applied here. In case of suburban transport, there are only two classes, first class and second class and both are not air conditioned. Hence, the entire suburban transport has been taken as non-AC class. However, the probable in-vehicle energy consumption of suburban AC coaches was also estimated to determine the in-vehicle energy use of AC coaches of suburban transport in the future.

In-Vehicle Electricity Consumption Per PKM of Non-Suburban Coaches

Using the specifications of Self-Generating coach for AC and non-AC passenger carriages, the average electrical energy consumed for in-vehicle use in AC coaches per PKM was estimated to be 9.80 watt-hours; the corresponding figure for non-AC coaches was estimated to be 0.74 watt-hours. Although, the ratio of average in-vehicle electrical energy consumption between AC and non-AC coach was estimated to be 7.41, the ratio of average electrical energy consumption per PKM for in-vehicle use between AC and non-AC coaches was found to be 13.24. The difference is because average number of passengers travelling in AC coaches is lower than that in non-AC coaches. AC coaches such as 1AC, 2AC and 3AC have 24 seats, 46 seats and 64 seats per coach, respectively, whereas non-AC coaches such as sleeper and seating class have 72 seats and 108 seats per coach, respectively. There is no upper limit on the number of passengers who can travel in general compartments.

In-Vehicle Electrical Energy Consumption of Passenger Carriages in Suburban Transport

Unlike the non-suburban passenger carriages, suburban passenger carriages have only non-AC coaches at present. Using the specifications of Self-Generating coach for non-AC passenger carriages, the in-vehicle energy consumption for on PKM in suburban travel was estimated to be 0.70 watt-hours. The reason for the slight difference may be the regular feature of overcrowding in suburban trains, which decreases the per capita consumption of in-vehicle electrical energy.

The values estimated for in-vehicle energy consumption are based on the electrical equipment fitted in the passenger carriages at present and the same values will be retained for 2015–16, the first year of predicting future electrical energy requirement.

8.3.4.2 Second Step: Electricity Energy Consumed for Only Hauling One GTKM of Rail Transport in General and Hauling One TKM of Freight Transport

Once the total energy spent on in-vehicle consumption was found, the remaining energy was spent on hauling of passenger and freight trains. Given the GTKM hauled for passenger and freight trains in 2010–11, the energy consumed for hauling one GTKM, irrespective of whether it is passenger or freight transport, was estimated as 12.58 watt-hours in 2010–11. The electricity consumed for hauling of one TKM of freight transport was estimated to be 22.74 watt-hour. The ratio GTKM/TKM was found to be 1.81. That is to haul one TKM of freight traffic; IR hauled 1.81 tonnes of payload plus tare load.

8.3.4.3 Third Step: Calculation of Electrical Energy Consumed Only for the Hauling Part of Non-Suburban Passenger Transport

Some basic assumptions are to be made to calculate the electrical energy consumed for the hauling part of non-suburban passenger transport. The first assumption is that electric traction was applied across AC and non-AC coaches of non-suburban transport in the same proportion. However, in practice, trains that are hauled by electric traction may contain more AC coaches than the trains that run on diesel traction. Moreover, there are trains that have only AC coaches such as Shatabdi and Rajdhani, which may run completely on electrified route, thereby disproportionately, add more AC coaches to the electric hauling. However, such small variations cannot be accounted and hence not accounted here. The second assumption is that the total weight of the coach would be same irrespective of whether it is an AC coach or non-AC coach of non-suburban trains. The passenger facilities (such as sleepers, seating arrangement, electric equipment and structure for housekeeping) which can add to the tare weight are more in AC coaches compared to non-AC coaches of non-suburban transport. But, this would get reasonably compensated with high payload (more passengers per coach) in non-AC coaches. Any variation in total weight between AC coach and non-AC coach was not accounted here.

Based on these assumptions and as well as after deducting the in-vehicle energy consumed by passenger coaches, the energy consumption for hauling one PKM in AC coaches and one PKM in non-AC coaches were estimated to be 14.07 and 7.83 watt-hours. The ratio GTKM/PKM was found to be 1.12 and 0.62 for AC classes and non-AC classes, respectively. To carry one passenger with his luggage for one km, AC classes of non-suburban transport and non-AC classes of non-suburban transport hauled 1120 kg and 620 kg, respectively. Such an enormous difference stems primarily from the number of passengers travelling in the respective coaches.

8.3.4.4 Fourth Step: Calculation of Electrical Energy Consumed Only for the Hauling Part of Suburban Passenger Transport

After deducting the in-vehicle energy consumed by passenger coaches, the energy consumption for hauling one PKM in non-AC coaches of suburban transport was estimated to be 3.63 watt-hours in 2010–11. The ratio GTKM/PKM was found to be 0.29. That is to haul one passenger for 1 km of suburban passenger traffic; IR hauled 290 kg of payload plus tare load. This low value was essentially due to crush load of suburban trains.

8.3.4.5 Fifth Step: Calculation of Electrical Energy Consumed for High-Speed Rail Transport

The total energy consumed to haul one PKM of High-Speed Rail with Shinkansen 700 series was 29 watt-hours. Shinkansen 700 series run at a maximum speed and service speed of 300 and 270 kmph, respectively (Network Rail n.d). The total energy consumed for one PKM of AC classes of non-suburban passenger transport in 2010–11 was 24 watt-hours, which is about 82% of the energy required for HSR. Although the energy required to haul the train at an average speed of 270 kmph was much larger than to haul conventional trains of IR at 60 kmph, the reduced travel time in HSR and hence the resultant reduced in-vehicle energy consumption nullifies the higher energy requirement for hauling.

8.3.5 Summary of the Electric Energy Consumption Data in Terms of PKM and TKM for 2010–11

Based on the estimates made in Sect. 8.3.2, Table 8.6 summarizes the findings that were made on electric energy consumption of IR in 2010–11. The introduction of

Table 8.6 Summary of electric energy consumption data in terms of PKM and TKM for 2010–11

Type of transport	Energy required for one PKM (watt-hours)		
	For hauling alone	For in-vehicle use	Total
Suburban non-AC class	3.63	0.70	4.33
Suburban AC class (anticipated)	3.63	2.52	6.15
Non-suburban AC class	14.12	9.80	23.92
Non-suburban non AC class	7.85	0.74	8.59
Type of transport	Energy required for one TKM (watt-hours)		
	For hauling alone	For in-vehicle use	Total
Freight	22.74	0	22.74

Source Author's calculation

AC class coaches in suburban rail services would increase the consumption only by about 40%. The growth of non-suburban AC class passenger transport in preference to non-suburban non-AC class would increase the electric energy consumption of IR tremendously.

8.4 Expected Future Trends and Assumptions Made Therein on Electrical Energy Consumption

The energy required per unit of rail transport as shown in Table 8.6 was for 2010–11. However, the energy requirement for future rail passenger (suburban and non-suburban) and freight traffic would be different due to several reasons like improvement in locomotives, in-vehicle electrical equipment, reduced weight of passenger coaches and freight wagons, more AC coaches in use over non-AC coaches for passenger travel, etc. So, it is imperative to outline the expected trends in energy consumption and thereby arrive at how the energy consumption per unit of transport would change in the years to come, as below.

8.4.1 Future Trends of Electrical Energy Consumption in Non-Suburban Passenger Transport

The average electricity consumed per thousand GTKM of electric hauling decreased from 19.7 units in 2003–04 to 18.90 in 2014–15 for passenger transport (across suburban and non-suburban transport) resulting in a CAGR of −0.38% between 2003–04 and 2014–15. In other words, the percentage change in energy consumption between 2003–04 and 2014–15 per unit of passenger transport has been −4.06 (MoR n.d.c; MoR n.d.j).

The energy efficiency improvement in electric traction system (such as change from 1500 V DC to 25 kV AC in Mumbai suburban system) and increased efficiency of the electric locomotives such as regenerative braking system and energy efficiency improvement in the electric equipment such as lights, fans and air conditioners in passenger carriages would have decreased the average electricity energy consumed per thousand GTKM. On the other hand, the increase in the number of AC coaches would have increased the average electric energy consumed per thousand GTKM of passenger transport, given the fact that the share of AC classes in the total passenger transport increased from 5.75% in 2003–04 to 8.77% in 2014–15.

IR planned to manufacture 12,000 HP Electric Locomotives through Joint Venture by setting up of Electric Locomotive Factory at Madhepura, Bihar. This would achieve about 87% energy efficiency, saving about 13% of the fuel consumed now (Indian Railways Green Energy Initiatives 2015). These locos will also reduce CO_2 emission of 350–500 tonne per loco per annum due to regeneration because of IGBT technology. Talgo passenger trains are under testing and would reduce the energy consumption of non-suburban trains by about 30% due to its reduced weight and inalterability against atmospheric agents (Biswas April 23 2016). To begin with, 15 train sets (each train set will have 21 passenger carriages) are to be introduced to IR non-suburban passenger services.

All the above measures may reduce the hauling energy consumption of passenger transport by about 43%, which corresponds to a CAGR of electrical energy consumption −3.68% between 2015–16 and 2030–31. There will be some more measures in the future that would reduce the energy consumption of hauling. However, it will take about 15–20 years to extend the changes completely to the entire network of IR. Hence, the assumption on the reduction of energy consumption is based only on the present conditions. In case of in-vehicle consumption in passenger trains of AC and non-AC coaches, the scope is very limited, as most of the energy-saving measures were already introduced. Hence, it is assumed here that the CAGR of in-vehicle electrical energy consumption would be −1.0% between 2015–16 and 2030–31 (overall energy reduction of 14% between 2015–16 and 2030–31).

8.4.2 Future Trends of Electrical Energy Consumption in Suburban Passenger Transport

New EMU coaches manufactured in Rail Coach Factory at Kanchrapara, West Bengal, which have regenerative capability would be deployed for suburban rail system (Indian Railways Green Energy Initiatives 2015). With the phased introduction of Kanchrapara EMU coaches, the energy consumption for hauling one PKM of suburban transport would decrease at a CAGR of 1% between 2015–16 and 2030–31 (14% between 2015–16 and 2030–31).

AC trains for suburban rail system have already been manufactured and would be introduced in Mumbai suburban rail system (Bhutada April 6 2016). On the assumption that EMU coaches will be introduced in 2017–18, if 7% of the existing EMU coaches (which do not have AC facility) are replaced with EMU coaches with AC facility every year, in 2030–31, the entire suburban rail system may have EMU carriages with AC facility. The energy consumption for in-vehicle use for EMU AC coaches would be 2.52 watt-hour per PKM. However, here no further reduction in the in-vehicle energy consumption of AC and non-AC EMU coaches is assumed, as the values are already low and cannot decrease further.

8.4.3 *Future Trends of Electrical Energy Consumption in Freight Transport*

The average electricity consumed per thousand GTKM of electric hauling decreased from 8.32 units in 2003–04 to 6.86 in 2014–15 for freight transport resulting in a CAGR of −1.74% between 2003–04 and 2014–15. In other words, the percentage change in energy consumption between 2003–04 and 2014–15 per unit of freight transport has been −17.55 at a CAGR of −1.74% (MoR n.d.c; MoR n.d.j). However, the future trends would depend on energy saving measures in the realm of rail freight transport. With lighter wagons and more payload per wagon, the electrical energy consumed for every TKM of freight transport will decrease. The existing wagons carry a payload of about 52–57 tonnes. IR has been acquiring and replacing the existing wagons which has a payload to tare ratio of 2.7 with stainless steel wagons payload to tare ratio of 3.4, which would help IR to transport additional 2 tonnes per wagon, which improves the energy required to haul one TKM of freight marginally, if not substantially (Business Standard November 25, 2015). The 9000 HP electric locomotives with regenerative braking features and equipped with IGBT Propulsion technology will be used on Western Dedicated Freight Corridor (Indian Railways Green Energy Initiatives 2015). IR also planned Double Stack Container Train operation for WDFC which would increase throughput up to 360 TEU per train (PIB Aug 7 2015). Taking all these into consideration, it is assumed here that the CAGR of energy consumed per unit of freight traffic would be −1.5% between 2016–17 and 2030–31, which corresponds to 20.3% reduction in energy consumption between 2016–17 and 2030–31.

8.4.4 *Summary of CAGR of Energy Consumption Per Unit of Traffic Between 2015–16 and 2030–31*

Based on the discussion in Sects. 8.4.1–8.4.3, Table 8.7 presents a summary of the CAGR of energy consumption per PKM for non-suburban and suburban passenger traffic, and per TKM for freight traffic up to 2030–31.

8.5 Estimating Future Electrical Traction in Rail Passenger and Freight Transport

Estimating future electrical traction in rail passenger and freight transport is to be based on running track length electrified than route length electrified, as double lines carry traffic which is four times that of single line. Moreover, the data on share of running track electrified give a broad picture of electrified hauling in IR. However, it would be incorrect to take % share of electrified running track as the

Table 8.7 Summary of CAGR of energy consumption per unit of traffic between 2015–16 and 2030–31

Type of transport	Category	Class	Energy consumption purpose	CAGR between 2015–16 and 2030–31 (%)
Passenger	Non-suburban	AC Classes	Hauling alone	−3.68
			In-vehicle use	−1.00
		Non-AC Classes	Hauling alone	−3.68
			In-vehicle use	−1.00
	Suburban	AC Classes	Hauling alone	−1.00
			In-vehicle use	0
		Non-AC Classes	Hauling alone	−1.00
			In-vehicle use	0
Freight			Hauling alone	−1.50

Source Author's calculation

proxy variable for % share of electric traction of passenger and freight traffic. When the entire rail network between the origin-destination is electrified, there are no issues for the use of electric locomotives for passenger and freight trains. However, when part of the rail network between the origin-destination is electrified, there are issues in running train with electric locomotives. In such a case, for all practical reasons, the trains may be operated on diesel traction to avoid delay and to overcome operational cumbersomeness of changing locomotives at various points.

So, the better solution would be to get the actual data on electric hauling both passenger (suburban and non-suburban) and freight transport. However, there is no data consistently available on year-on-year basis on how much PKM of suburban and non-suburban and rail freight transport has been hauled by electric traction. The data are available intermittently for few years (2000–01, 2007–08, 2008–09, 2010–11 and 2013–14). From this and from other data, the relationship between percentage of electrified running tracks and suburban passenger transport, percentage of electrified running tracks and non-suburban passenger transport and percentage of electrified running tracks and freight transport.

8.5.1 *Establishing the Relationship Between Percentage of Electrified Running Tracks and Suburban Passenger Transport*

In case of suburban transport, it is assumed that the entire suburban rail network has been operating on electrified network as the suburban train systems of major cities are on run by EMUs. However, DMU and DHMU have also been deployed for the suburban train services where the network is not electrified. From the data of Indian Railways, the EMUs constitute about 10% of the total suburban trains. However,

DMU/DHMU run in some suburban segments mainly because of lack of electrification in some segments. DMU/DHMU would vanish sooner compared to non-suburban services if the railway network in which suburban services ply is completely electrified. Hence here, it is assumed that the change in electric hauling would be 0.75% per year from 2015–16 onwards and by 2030–31, 99.15% of the suburban rail system would be operated by electric traction.

8.5.2 Establishing the Relationship Between Percentage of Electrified Running Tracks and Percentage of Electrified Hauling of Non-Suburban Passenger Transport

Based on the data available on percentage of electrified running track and the percentage of passenger transport hauled by electric traction for 2000–01 and 2007–08 to 2010–11, after some adjustments for suburban passenger transport, the conversion factor was found to be varying between 1.15 and 1.27. The lower value of 1.15 was taken to determine the electric hauling of non-suburban passenger transport up to 2030–31, which means if 60% of the total running track is electrified, Indian Railways would haul 69% of the total non-suburban passenger transport by electric traction.

8.5.3 Establishing the Relationship Between Percentage of Electrified Running Tracks and Percentage of Electrified Hauling of Freight Transport

Based on the data available on percentage of electrified running track and the percentage of freight transport hauled by electric traction for 2000–01, 2007–08 to 2010–11 and 2013–14, the conversion factor was found to be varying between 1.48 and 1.75. The lower value of 1.48 was taken to determine the electric hauling of freight transport up to 2030–31, which means if 60% of the total running track is electrified, Indian Railways would haul 89% of the total freight transport by electric traction.

8.5.4 Expected Growth of Electrification in Rail Network and Electric Traction

The overall running track length has been increasing at 638 km per year between 2000–01 and 2014–15 in the conventional rail. However, a new impetus has been given for additional tracks by aiming to lay 7 km of track every day in 2016–17 (Hindustan Times Feb 28 2016), Hence, it assumed here that the running track length would be increased at a rate of 1000 km per year between 2014–15 and 2030–31 without electrification. The electrified running track length has been increasing at 936 km per year between 2000–01 and 2014–15 in the conventional rail. However, IR planned to electrify tracks at a rate of about 1500–2000 km per year till 2020–21 and 2000 km per year thereafter in the conventional rail. In addition, the Ministry of Power also come out with a plan on electrification of 35,000 km of railway lines in 3 years from its own resources to save forex (Economic Times Mar 03 2016). In this backdrop, it is assumed here that the electrification may be carried out at a rate of 1500 running track km per year between 2014–15 and 2030–31, which will increase the total electrified running track by 24,000 km by 2030–31.

It is assumed here that 1000 km of non-electrified tracks would be added per year between 2014–15 and 2030–31 and the existing running tracks are electrified at a rate of 1500 km per year between 2014–15 and 2030–31. As a result, the nonelectrified track length would be 8000 km lower than that of 2014–15. The WDFC and EDFC would be expected to be operationalized in 2020–21 and HSR between Ahmedabad–Mumbai would be operationalized in 2024–25. This would bring a substantial increase in the electric traction of freight transport. Other DFCs and HSR projects, which are in feasibility stage, as explained in Sect. 8.2.1 would be expected to be commissioned between 2027–28 and 2030–31. The Greenfield projects of DFC and HSR also are developed as electrified tracks. Every running track kilometre of DFC would provide three times the capacity of that of every running track kilometre of conventional rail track as DFC trains would be operated at an average speed of 75 kmph compared to 25 kmph of conventional trains. Hence, a weightage of 3 is given for DFC in freight traffic. Two-third of the existing rail network has been used for passenger transport and one-third is used for freight transport. With the implementation of DFC, the entire capacity of the existing network will be available for passenger transport. That is every 3 km of running track of DFC would provide additional capacity of 1 km of running track for rail passenger transport. For a given route length or running track length, the carrying capacity of HSR is about five times that of conventional network. A double line conventional rail can ply six trains in both directions (three in one direction with a headway of 20 min) and a corresponding double line HSR can ply 30 trains in both directions (15 in one direction with a headway of 4 min). If the capacity of HSR train and conventional train are assumed to be equal, the carrying capacity of an HSR line is about five times that of conventional rail, for a given length.

Based on the above discussion, it was estimated that IR will have 92,650 km of effective electrified conventional rail running track against the total running track length of 134,415 km for non-suburban passenger transport in 2030–31, which corresponds to 79% of the non-suburban passenger traffic hauled by electric traction. This estimation could go wrong only if Indian government comes up with a massive plan of large-scale electrification of existing rail network and execute the same in time-bound manner. The Ministry of Power wanted to electrify 30,000 km of rail network within 3 years to reduce oil import, reduce fuel expenses (the cost of electric hauling is about 45% lower than the cost of diesel hauling) and save foreign exchange (Economic Times Mar 03 2016).

It was estimated that IR will have 134,242 km of effective electrified conventional rail running track against the total running track length of 176,007 km for non-suburban passenger transport in 2030–31, which corresponds to 100% of the freight traffic hauled by electric traction. This is mainly because the freight trains have been running mainly between industrial centres, ports, power plants and coal mines, whereas passenger trains reach to even smaller towns. In 2014–15, IR had 41,038 km of electrified conventional rail running track against the total running track length of 90,803 km for non-suburban passenger transport.

8.6 Rail Passenger Transport Forecast and Electrical Energy Requirement up to 2030–31

The expected rail passenger transport forecast up to 2030–31 was outlined in Sect. 8.2 for the three scenarios of optimistic, realistic and pessimistic GDP CAGR for passenger transport. However, the cumulative rail passenger forecast scenarios were outlined only for 2019–20, 2024–25 and 2030–31, without a detailed year-on-year growth. To understand how the rail passenger transport demand would grow in non-suburban and suburban rail systems and the classes within, it is imperative to look at the growth of these passenger segments in the recent past.

8.6.1 Rail Passenger Transport Growth Between 2003–04 and 2014–15

The non-suburban passenger transport segment had grown at a CAGR of 7.59%, whereas the suburban passenger transport segment had grown at a CAGR of 4.25% between 2003–04 and 2014–15. The capacity of suburban rail systems has been utilized to the maximum extent which is evident from crush load under which suburban passengers have been travelling and as a result, the suburban passenger segment has been moving to personal transport (especially two-wheelers), paratransit and bus services. There is a very limited scope for suburban rail infrastructure to expand further all along the existing routes as there is no space to create additional rail tracks. In routes where suburban rail systems are not available, the metro systems are developed to provide rail connectivity in cities. So, the suburban rail passenger transport operated by Indian Railways may not increase substantially in absolute numbers. However, the long distance intercity travel by rail would continue to increase and hence the share of suburban rail passenger transport in the overall rail passenger transport operated by Indian Railways would shrink in future.

Even within the non-suburban passenger segment, AC classes had grown at a CAGR of 11.26%, whereas non-AC classes had grown at a CAGR of 7.26% between 2003–04 and 2014–15. The tendency to travel by AC classes would increase further in the impending years. The share of AC classes in non-suburban passenger transport increased from 5.75% in 2003–04 to 8.77% in 2014–15 with no dip in any year. With rising income of people, people look for comfortable travel and AC coaches serve that purpose. On the other hand, IR increased the AC coaches in passenger trains and introduced plenty of exclusive AC trains since 2005. Even the trains meant for the so-called poorer sections of the society, such Garib Rath and Duronto were introduced as exclusive AC trains (Ramakrishnan July 17 2015). AC 3 tier breaks even at approximately 75% occupancy and it is the only coach which has given profits for IR (MoR 2015). Moreover, it has been preferred by the passengers also. In this backdrop, the AC class travel would increase substantially in the impending years.

8.6.2 Rail Passenger Transport Forecast up to 2030–31

It is not just enough to know the overall passenger forecast up to 2030–31. As the in-vehicle electricity consumption per PKM varies between suburban and non-suburban train travel as well as between AC and non-AC classes in both category, it is essential to forecast the passengers in each category up to 2030–31.

Based on the inferences mentioned in Sect. 8.6.1, the assumptions for the demand of rail passenger transport in suburban and non-suburban categories of AC and non-AC travel between 2015–6 and 2030–31 for the three scenarios were made. The logic behind choosing the CAGR for various time periods is explained as follows.

8.6.2.1 CAGR Between 2015–16 and 2019–20

The passenger transport in terms of Million PKM of non-suburban non-AC class travel decreased by 7055 between 2013–14 and 2014–15. The passenger transport in terms of Million PKM of suburban travel increased by 1516 between 2013–14 and 2014–15. On the other hand, passenger transport in terms of Million PKM of suburban travel increased by 12,317 between 2013–14 and 2014–15 and the overall passenger transport in terms of Million PKM increased by 6778. The trend is very clear that growth of non-suburban AC travel would eclipse other forms of travel. The expected growth in total passenger transport has been moderately low (6.30%) in this period and hence the CAGR between 2003–04 and 2014–15 of suburban travel and the AC classes of non-suburban travel have been taken as the CAGR for this period also and the rest is assigned to non-suburban non-AC travel.

8.6.2.2 CAGR Between 2020–21 and 2024–25

Although the capacity of suburban rail systems is almost saturated, with improvement in travel experience such as AC coaches and some innovative measures, the past trend of suburban travel may be retained. With the expected introduction of AC suburban coaches in suburban rail systems in 2017–18, the suburban travel may go at historical CAGR of 4.25%. The non-AC travel of non-suburban travel would also increase at a historical CAGR of 7.26%. The rest would go the AC non-suburban travel, expecting that the demand and supply of non-suburban AC travel would be very high between 2020–21 and 2024–25. As 3AC coaches attain break-even for IR even with 75% occupancy and 3AC class remain the preferred luxury class of rail passengers, 3AC coach manufacturing has been prioritized over the manufacturing of other coaches (MoR 2015). This is the period where the proliferation of semi-HSR trains across various routes and the introduction of first HSR between Ahmedabad and Mumbai are expected to happen and this would be able to provide the high demand of non-suburban AC class travel.

8.6.2.3 CAGR Between 2025–26 and 2030–31

The demand pattern between 2020–21 and 2024–25 would be retained for this period also. In this period, the consolidation of HSR and the introduction of some more routes on HSR are expected to happen. IR might have enough capacity by 2025–26, so that the passenger transport demand especially of non-suburban AC class travel would increase unabatedly. The above discussion is summarized in Table 8.8.

Based on the estimates made in the previous sections for passenger transport of both suburban and non-suburban transport and AC and non-AC classes, rail passenger transport forecast and the associated energy requirement between 2015–16 and 2030–31 are estimated. Tables 8.9 and 8.10 summarized rail passenger forecast and rail passenger forecast hauled by electric traction respectively up to 2030–31 for optimistic scenario. Tables 8.11 and 8.12 summarized rail passenger forecast and rail passenger forecast hauled by electric traction respectively up to 2030–31 for realistic scenario. Tables 8.13 and 8.14 summarized rail passenger forecast and rail passenger forecast hauled by electric traction respectively up to 2030–31 for pessimistic scenario.

Table 8.8 Assumed CAGR for passenger transport between 2015–16 and 2030–31 for three scenarios

Period	Description	Optimistic scenario	Realistic scenario	Pessimistic scenario
2015–16 to 2019–20	Total passenger transport	6.30	5.59	4.9
	Non-suburban AC	11.26	11.26	11.26
	Non-suburban non-AC	6.02	5.10	4.19
	Suburban	4.25	4.25	4.25
2020–21 to 2024–25	Total passenger transport	12.60	11.20	9.80
	Non-suburban AC	38.88	32.54	25.56
	Non-suburban non-AC	7.26	7.26	7.26
	Suburban	4.25	4.25	4.25
2025–26 to 2030–31	Total passenger transport	14.40	12.8	11.20
	Non-suburban AC	25.30	23.58	21.54
	Non-suburban non-AC	7.26	7.26	7.26
	Suburban	4.25	4.25	4.25

Source Author's calculation

8.6.3 Electrical Energy Requirement for Rail Passenger Transport up to 2030–31

Based on the summary of CAGR of energy consumption per unit of traffic between 2015–16 and 2030–31 mentioned in Table 8.7 and the rail passenger growth forecast estimated for the three scenarios as shown in Tables 8.9, 8.10, 8.11, 8.12, 8.13 and 8.14, the electrical energy requirement for rail passenger transport up to 2030–31 was estimated for all the three scenarios in Tables 8.15, 8.16 and 8.17.

8.7 Rail Freight Transport Forecast and Electrical Energy Requirement up to 2030–31

The expected rail freight transport forecast up to 2030–31 was outlined in Sect. 8.2 for the three scenarios of optimistic, realistic and pessimistic GDP CAGR. The rail freight data between 2003–04 and 2014–15 indicate that there is no consistent growth across the years. The trend of low growth in freight transport is expected to continue till IR improves the speed and quality of freight service. However, for that to happen, Greenfield infrastructures such as DFCs are to be operationalized. The WDFC and EDFC are expected to be operationalized in 2020. The freight transport growth remains modest till that time and then picks up. Taking these factors into consideration, the three scenarios of rail freight transport forecast and

Table 8.9 Rail passenger forecast up to 2030–31 for optimistic scenario

Year	Passenger Transport (Million PKM)							Share in passenger transport (%)				Remarks
	Non-suburban			Suburban			All total	Non-suburban		Suburban		
	AC	Non-AC	Total	AC	Non-AC	Total		AC	Non-AC	AC	Non-AC	
2014–15	100,616	894,799	995,415	0	151,775	151,775	1,147,190	8.77	78	0	13.23	Actual
2019–20	171,537	1,198,576	1,370,113	42,050	144,837	186,887	1,557,000	11.02	76.98	2.70	9.30	Forecast
CAGR[a]	11.26	6.02	6.6	NA	−0.93	4.25	6.3	NA	NA	NA	NA	
2024–25	886,289	1,701,589	2,587,878	138,073	92,049	230,122	2,818,000	31.45	60.38	4.9	3.27	Forecast
CAGR [b]	38.88	7.26	13.56	26.84	−8.67	4.25	12.6	NA	NA	NA	NA	
2030–31	3,430,514	2,591,083	6,021,597	295,403	0	295,403	6,317,000	54.31	41.02	4.68	0	Forecast
CAGR [c]	25.3	7.26	15.11	13.51	−100	4.25	14.4	NA	NA	NA	NA	
CAGR [d]	24.68	6.87	11.91	NA	−100	4.25	11.25	NA	NA	NA	NA	

[a]between 2014–15 and 2019–20 (%), [b]between 2020–21 and 2024–25 (%), [c]between 2025–26 and 2030–31 (%), [d]between 2014–15 and 2030–31 (%)
Source Author's forecast

Table 8.10 Rail passenger forecast hauled by electric traction up to 2030–31 for optimistic scenario

	Share of electric hauling (%)		Passenger transport by electric hauling (Million PKM)							Remarks
	Non-suburban	Suburban	Non-suburban			Suburban			All total	
Year			AC	Non-AC	Total	AC	Non-AC	Total		
2014–15	51.97	87.15	52,290	465,027	517,317	0	132,272	132,272	649,589	Actual
2019–20	58.26	90.9	99,937	698,290	798,228	38,223	131,657	169,880	968,108	Forecast
CAGR[a]	NA	NA	13.83	8.47	9.06	NA	−0.09	5.13	8.31	
2024–25	67.59	94.65	599,043	1,150,104	1,749,147	130,686	87,124	217,810	1,966,957	Forecast
CAGR[b]	NA	NA	43.07	10.49	16.99	27.87	−7.93	5.1	15.23	
2030–31	79.27	99.15	2,719,368	2,053,951	4,773,320	292,892	0	292,892	5,066,212	Forecast
CAGR[c]			28.68	10.15	18.21	14.4	−100	5.06	17.08	
CAGR[d]			28.01	9.73	14.9	NA	−100	5.09	13.7	

[a]between 2014–15 and 2019–20 (%), [b]between 2020–21 and 2024–25 (%), [c]between 2025–26 and 2030–31 (%), [d]between 2014–15 and 2030–31 (%),
Source Author's forecast

Table 8.11 Rail passenger forecast demand between up to 2030–31 for realistic scenario

CAGR of AC coach commuters in suburban transport (%)			7.50									Remarks
	Passenger Transport (Million PKM)							Share in passenger transport (%)				
Year	Non-suburban			Suburban			All total	Non-suburban		Suburban		
	AC	Non-AC	Total	AC	Non-AC	Total		AC	Non-AC	AC	Non-AC	
2014–15	100,616	894,799	995,415	0	151,775	151,775	1,147,190	8.77	78	0	13.23	Actual
2019–20	171,537	1,147,576	1,319,113	42,050	144,837	186,887	1,506,000	11.39	76.2	2.79	9.62	Forecast
CAGR[a]	11.26	5.1	5.79	NA	−0.93	4.25	5.59	NA	NA	NA	NA	
2024–25	701,693	1,629,185	2,330,878	138,073	92,049	230,122	2,561,000	27.4	63.62	5.39	3.59	Forecast
CAGR[b]	32.54	7.26	12.06	26.84	−8.67	4.25	11.2	NA	NA	NA	NA	
2030–31	2,499,767	2,480,830	4,980,597	295,403	0	295,403	5,276,000	47.38	47.02	5.6	0	Forecast
CAGR[c]	23.58	7.26	13.49	13.51	−100	4.25	12.8	NA	NA	NA	NA	
CAGR[d]	22.24	6.58	10.59	NA	−100	4.25	10.01	NA	NA	NA	NA	

[a]between 2014–15 and 2019–20 (%), [b]between 2020–21 and 2024–25 (%), [c]between 2025–26 and 2030–31 (%), [d]between 2014–15 and 2030–31 (%)
Source Author's forecast

Table 8.12 Rail passenger forecast hauled by electric traction up to 2030–31 for realistic scenario

Year	Share of electric hauling (%)		Passenger transport by electric hauling (Million PKM)							Remarks
	Non-suburban	Suburban	Non-suburban			Suburban				
			AC	Non-AC	Total	AC	Non-AC	Total	All total	
2014–15	51.97	87.15	52,290	465,027	517,317	0	132,272	132,272	649,589	Actual
2019–20	53.26	90.90	99,937	668,578	768,515	38,223	131,657	169,880	938,395	Forecast
CAGR[a]			13.83	7.53	8.24	NA	−0.09	5.13	7.63	
2024–25	67.59	94.65	474,274	1,101,166	1,575,440	130,686	87,124	217,810	1,793,250	Forecast
CAGR[b]			36.54	10.49	15.44	27.87	−7.93	5.1	13.83	
2030–31	79.27	99.15	1,981,565	1,966,554	3,948,119	292,892	0	292,892	4,241,011	Forecast
CAGR[c]			26.91	10.15	16.55	14.4	−100	5.06	15.43	
CAGR[d]			25.51	9.43	13.54	NA	−100	5.09	12.44	

[a]between 2014–15 and 2019–20 (%), [b]between 2020–21 and 2024–25 (%), [c]between 2025–26 and 2030–31 (%), [d]between 2014–15 and 2030–31 (%)

Source Author's forecast

Table 8.13 Rail passenger forecast demand between up to 2030–31 for pessimistic scenario

Year	Passenger Transport (Million PKM)							Share in passenger transport (%)				Remarks
	Non-suburban			Suburban			All total	Non-suburban		Suburban		
	AC	Non-AC	Total	AC	Non-AC	Total		AC	Non-AC	AC	Non-AC	
2014–15	100,616	894,799	995,415	0	151,775	151,775	1,147,190	8.77	78	0	13.23	Actual
2019–20	171,537	1,098,576	1,270,113	42,050	144,837	186,887	1,457,000	11.77	75.4	2.89	9.94	Forecast
CAGR[a]	11.26	4.19	4.99	NA	−0.93	4.25	4.9	NA	NA	NA	NA	
2024–25	535,256	1,559,622	2,094,878	138,073	92,049	230,122	2,325,000	23.02	67.08	5.94	3.96	Forecast
CAGR[b]	25.56	7.26	10.53	26.84	−8.67	4.25	9.8	NA	NA	NA	NA	
2030–31	1,725,693	2,374,904	4,100,597	295,403	0	295,403	4,396,000	39.26	54.02	6.72	0	Forecast
CAGR[c]	21.54	7.26	11.84	13.51	−100	4.25	11.2	NA	NA	NA	NA	
CAGR[d]	19.44	6.29	9.25	NA	−100	4.25	8.76	NA	NA	NA	NA	

[a] between 2014–15 and 2019–20 (%), [b]between 2020–21 and 2024–25 (%), [c]between 2025–26 and 2030–31 (%), [d]between 2014–15 and 2030–31 (%)
Source Author's forecast

Table 8.14 Rail passenger forecast hauled by electric traction up to 2030–31 for pessimistic scenario

Year	Share of electric hauling (%)		Passenger transport by electric hauling (Million PKM)							Remarks
	Non-suburban	Suburban	Non-suburban			Suburban			All total	
			AC	Non-AC	Total	AC	Non-AC	Total		
2014–15	51.97	87.15	52,290	465,027	517,317	0	132,272	132,272	649,589	Actual
2019–20	58.26	90.90	99,937	640,030	739,968	38,223	131,657	169,880	909,848	Forecast
CAGR[a]			13.83	6.6	7.42	NA	−0.09	5.13	6.97	
2024–25	67.59	94.65	361,780	1,054,149	1,415,928	130,686	87,124	217,810	1,633,738	Forecast
CAGR[b]			29.34	10.49	13.86	27.87	−7.93	5.1	12.42	
2030–31	79.27	99.15	1,367,957	1,882,586	3,250,543	292,892	0	292,892	3,543,435	Forecast
CAGR[c]			24.82	10.15	14.86	14.4	−100	5.06	13.77	
CAGR[d]			22.63	9.13	12.17	NA	−100	5.09	11.19	

[a]between 2014–15 and 2019–20 (%), [b]between 2020–21 and 2024–25 (%), [c]between 2025–26 and 2030–31 (%), [d]between 2014–15 and 2030–31 (%)

Source Author's forecast

Table 8.15 Electricity requirement for electric traction of rail passenger transport up to 2030–31 for optimistic scenario

Year	Electrical energy requirement for passenger transport (MU)												
	Non-suburban						Suburban						All total
	AC			Non-AC			AC			Non-AC			
	Hauling	In-vehicle use	Total	Hauling	In-vehicle use	Total	Hauling	In-vehicle use	Total	Hauling	In-vehicle use	Total	
2019–20	1066	940	2006	4141	489	4530	131	97	228	453	92	545	7409
CAGR[a]	9.9	12.92	11.25	4.72	7.29	4.98	NA	NA	NA	−0.94	−0.22	−0.82	6.62
2024–25	5302	5367	10,669	5647	748	6395	430	331	761	287	61	348	18,173
CAGR[b]	37.83	41.68	39.69	6.4	8.87		26.84	27.82	27.26	−8.72	−7.89	-8.58	19.66
2030–31	19,226	22,897	42,123	8051	1212	9263	911	741	1652	0	0	0	53,038
CAGR[c]	23.95	27.35	25.72	6.09	8.38		13.33	14.38	13.79	−100	−100	−100	19.54
CAGR[d]	23.4	26.81	25.06	12.77	−6.05	22.85	−8.28	NA	NA	NA	−100	−100	32.78

[a]between 2014–15 and 2019–20 (%), [b]between 2020–21 and 2024–25 (%), [c]between 2025–26 and 2030–31 (%), [d]between 2014–15 and 2030–31 (%)
Source Author's forecast

Table 8.16 Electricity requirement for electric traction of rail passenger transport up to 2030–31 for realistic scenario

Year	Electrical energy consumed for passenger transport (MU)												
	Non-suburban						Suburban						All total
	AC			Non-AC			AC			Non-AC			
	Hauling	In-vehicle use	Total	Hauling	In-vehicle use	Total	Hauling	In-vehicle use	Total	Hauling	In-vehicle use	Total	
2019–20	1066	940	2006	3965	468	4433	131	97	228	453	92	545	7212
CAGR[a]	9.9	12.92	11.25	3.82	6.35	4.07	NA	NA	NA	−0.94	−0.22	−0.82	6.05
2024–25	4197	4249	8446	5407	716	6123	430	331	761	287	61	348	15,678
CAGR[b]	31.53	35.22	33.31	6.4	8.88		26.84	27.82	27.26	−8.72	−7.89	−8.58	16.8
2030–31	14,010	16,685	30,695	7709	1160	8869	911	741	1652	0	0	0	41,216
CAGR[c]	22.25	25.6	23.99	6.09	8.37		13.33	14.38	13.79	−100	−100	−100	17.48
CAGR[d]	20.98	24.33	22.61	12.46	−6.3	22.52	−8.28	NA	NA	NA	−100	−100	30.71

[a]between 2014–15 and 2019–20 (%), [b]between 2020–21 and 2024–25 (%), [c]between 2025–26 and 2030–31 (%), [d]between 2014–15 and 2030–31 (%)

Source Author's forecast

Table 8.17 Electricity requirement for electric traction of rail passenger transport up to 2030–31 for pessimistic scenario

Year	Electrical energy consumed for passenger transport (MU)												
	Non-suburban						Suburban						All total
	AC			Non-AC			AC			Non-AC			
	Hauling	In-vehicle use	Total	Hauling	In-vehicle use	Total	Hauling	In-vehicle use	Total	Hauling	In-vehicle use	Total	
2019–20	1066	940	2006	3795	448	4243	131	97	228	453	92	545	7022
CAGR[a]	9.9	12.92	11.25	2.91	5.43	3.16	NA	NA	NA	−0.94	−0.22	−0.82	5.48
2024–25	3202	3242	6444	5176	685	5861	430	331	761	287	61	348	13,414
CAGR[b]	24.6	28.1	25.29	6.4	8.86		26.84	27.82	27.26	−8.72	−7.89	−8.58	13.82
2030–31	9671	11,518	21,189	7380	1111	8491	911	741	1652	0	0	0	31,332
CAGR[c]	20.23	23.53	21.94	6.09	8.39		13.33	14.38	13.79	−100	−100	−100	15.19
CAGR[d]	18.21	21.48	19.8	12.16	−6.56	22.19	−8.28	NA	NA	NA	−100	−100	28.49

[a]between 2014–15 and 2019–20 (%), [b]between 2020–21 and 2024–25 (%), [c]between 2025–26 and 2030–31 (%), [d]between 2014–15 and 2030–31 (%)
Source: Author's forecast

Table 8.18 Rail freight transport forecast and energy requirement up to 2030–31 for optimistic scenario

CAGR of energy per TKM between 2015–16 and 2013–31				−1.50	
Year	Freight transport (Million TKM)	Share of electric hauling (%)	Passenger transport by electric hauling (Million PKM)	Electric energy required for traction (Watt-hour/TKM)	Electrical energy consumed for freight transport (MU)
2019–20	850,000	74.98	637,330	21.4	13,639
CAGR[a]	4.51	NA	6.93	NA	5.64
2024–25	2,260,000	93.39	2,110,614	19.84	41,875
CAGR[b]	21.6	NA	27.06	NA	25.15
2030–31	7,637,000	100	7,637,000	18.12	138,382
CAGR[c]	22.5	NA	23.9	NA	22.05
CAGR[d]	16.3	NA	19.26	NA	17.58

[a]Between 2014–15 and 2019–20, [b]Between 2020–21 and 2024–25, [c]Between 2025–26 and 2030–31 [d]Between 2014–15 and 2030–31
Source Author's forecast

the corresponding electrical energy requirement up to 2030–31 were developed and is shown in Tables 8.18, 8.19 and 8.20.

8.8 Forecast of Overall Electrical Energy Requirement of IR up to 2030–31

The share of non-transport electrical energy in the overall electrical energy consumed by IR was 17.1% in 2007–08 and reduced to 13.72% in 2014–15 with various energy saving measures and there is no further scope for IR to reduce its non-transport electrical energy requirement. Hence, it is assumed here that the non-transport electrical energy required for IR would be at about 15% of the total electrical energy requirement, which corresponds to 17.65% of the electrical energy required for transport. The overall rail transport electrical energy requirement up to 2030–31 was arrived by adding the electrical energy requirement for passenger and freight transport for each scenario and then 17.65% of the electrical energy required for rail transport as the non-transport component. Tables 8.21 and 8.22 show the overall energy requirement of IR up to 2030–31 for optimistic, realistic and pessimistic scenarios, respectively.

The key inferences from Tables 8.21, 8.22 and 8.23 on overall electrical energy requirement of IR up to 2030–31 are as follows (Table 8.24, 8.25 and 8.26).

Table 8.19 Rail freight transport forecast and energy requirement up to 2030–31 for realistic scenario

CAGR of energy per TKM between 2015–16 and 2013–31				−1.5	
Year	Freight transport (Million TKM)	Share of electric hauling (%)	Passenger transport by electric hauling (Million PKM)	Electric energy for traction (Watt-hour)	Electrical energy consumed for freight transport (MU)
2019–20	830,000	74.98	622,334	21.4	13,318
CAGR[a]	4.02	NA	6.42	NA	5.14
2024–25	1,997,000	93.39	1,864,998	19.84	37,002
CAGR[b]	19.2	NA	24.55	NA	22.68
2030–31	5,963,000	100	5,963,000	18.12	108,050
CAGR[c]	20.00	NA	21.38	NA	19.55
CAGR[d]	14.52	NA	17.43	NA	15.78

[a]Between 2014–15 and 2019–20, [b]Between 2020–21 and 2024–25, [c]Between 2025–26 and 2030–31, [d]Between 2014–15 and 203–31

Source Author's forecast

Table 8.20 Rail freight transport forecast and energy requirement up to 2030–31 for pessimistic scenario

Year	CAGR of energy per TKM between 2015–16 and 2013–31			−1.5	
	Freight transport (Million TKM)	Share of electric hauling (%)	Freight transport by electric hauling (Million PKM)	Electric energy for traction (Watt-hour)	Electrical energy consumed for freight transport (MU)
2019–20	810,000	74.98	607,338	21.4	12,997
CAGR[a]	3.51	NA	5.9	NA	18.37
2024–25	1,761,000	93.39	1,644,598	19.84	32,629
CAGR[b]	16.8	NA	22.05	NA	20.21
2030–31	4,634,000	100	4,634,000	18.12	83,968
CAGR[c]	17.50	NA	18.85	NA	17.06
CAGR[d]	12.73	NA	15.6	NA	18.45

[a]Between 2014–15 and 2019–20, [b]Between 2020–21 and 2024–25, [c]Between 2025–26 and 2030–31, [d]Between 2014–15 and 2030–31

Source Author's forecast

Table 8.21 Overall electrical energy requirement of IR up to 2030–31 for the optimistic scenario

	Energy requirement (MU)					Share (%)			
	Transport					Transport			Non- transport
Year	Passenger	Freight	Total	Non-transport	All	Passenger	Freight	Total	
2014–15	5377	10,368	15,745	2503	18,248	29.47	56.82	86.28	13.72
2019–20	7409	13,639	21,048	3715	24,763	29.92	55.08	85.00	15.00
CAGR[a]	6.62	5.64	5.98	8.22	6.3	NA	NA	NA	NA
2024–25	18,173	41,875	60,048	10,598	70,646	25.72	59.27	85.00	15.00
CAGR[b]	19.66	25.15	23.33	23.33	23.33	NA	NA	NA	NA
2030–31	53,038	138,382	191,420	33,786	225,206	23.55	61.45	85.00	15.00
CAGR[c]	19.54	22.05	21.32	21.32	21.32	NA	NA	NA	NA
CAGR[d]	32.78	17.58	16.9	17.66	17.01	NA	NA	NA	NA

[a]Between 2014–15 and 2019–20, [b]Between 2020–21 and 2024–25, [c]Between 2025–26 and 2030–31, [d]Between 2014–15 and 2030–31

Source Author's forecast

Table 8.22 Overall electrical energy requirement of IR up to 2030–31 for the realistic scenario

	Energy requirement (MU)					Share (%)			
	Transport			Non-transport	All total	Transport			Non-transport
Year	Passenger	Freight	Total			Passenger	Freight	Total	
2014–15	5377	10,368	15,745	2503	18,248	29.47	56.82	86.28	13.72
2019–20	7212	13,318	20,530	3624	24,154	29.86	55.14	85.00	15.00
CAGR[a]	6.05	5.14	5.45	7.68	5.77	NA	NA	NA	NA
2024–25	15,678	37,002	52,680	9298	61,978	25.3	59.7	85.00	15.00
CAGR[b]	16.8	22.68	20.74	20.74	20.74	NA	NA	NA	NA
2030–31	41,216	108,050	149,266	26,345	175,611	23.47	61.53	85.00	15.00
CAGR[c]	17.48	19.55	18.96	18.96	18.96	NA	NA	NA	NA
CAGR[d]	30.71	15.78	15.09	15.85	15.2	NA	NA	NA	NA

[a]Between 2014–15 and 2019–20, [b]Between 2020–21 and 2024–25, [c]Between 2025–26 and 2030–31, [d]Between 2014–15 and 2030–31

Source Author's forecast

Table 8.23 Overall electrical energy requirement of IR up to 2030–31 for the pessimistic scenario

	Energy requirement (MU)					Share (%)			Non-transport
	Transport					Transport			
Year	Passenger	Freight	Total	Non-transport	All total	Passenger	Freight	Total	
2014–15	5377	10,368	15,745	2503	18,248	29.47	56.82	86.28	13.72
2019–20	7022	12,997	20,019	3533	23,552	29.81	55.18	85.00	15.00
CAGR[a]	5.48	18.37	4.92	7.14	5.24	NA	NA	NA	NA
2024–25	13,414	32,629	46,043	8127	54,170	24.76	60.23	85.00	15.00
CAGR[b]	13.82	20.21	18.13	18.13	18.13	NA	NA	NA	NA
2030–31	31,332	83,968	115,300	20,350	135,650	23.1	61.9	85.00	15.00
CAGR[c]	15.19	17.06	16.53	16.53	16.53	NA	NA	NA	NA
CAGR[d]	23.49	18.45	13.25	13.99	13.36	NA	NA	NA	NA

Notes [a]Between 2014–15 and 2019–20, [b]Between 2020–21 and 2024–25, [c]Between 2025–26 and 2030–31, [d]Between 2014–15 and 2030–31

Source Author's forecast

Table 8.24 CO_2 equivalent emissions of passenger and freight transport up to 2030–31 for optimistic scenario

	CO_2 emissions (giga gram)							Share (%)			
	Passenger transport			Freight transport							
Year	Electricity	Diesel	Total	Electricity	Diesel	Total	All Total	Passenger	Freight	Electricity	Diesel
2014–15	5732	3539	9271	8903	4563	13,466	22,737	40.77	59.23	64.37	35.63
2019–20	8327	4028	12,355	11,832	4041	15.873	28,228	43.77	56.23	71.42	28.58
CAGR[a]	7.75	2.62	5.91	5.85	-2.40	3.34	4.42	NA	NA	NA	NA
2024–25	20,230	6634	26,364	35,430	4538	39,968	66,831	40.2	59.8	83.28	16.72
CAGR[b]	19.43	10.49	16.81	24.53	2.35	20.28	18.81	NA	NA	NA	NA
2030–31	59,232	10,829	70,051	119,671	5537	125,208	195,269	35.88	64.12	91.62	8.38
CAGR[c]	19.61	8.51	17.32	22.49	3.37	20.96	19.57	NA	NA	NA	NA
CAGR[d]	15.72	7.24	13.47	17.63	1.22	14.95	14.38	NA	NA	NA	NA

Notes [a]Between 2014–15 and 2019–20, [b]Between 2020–21 and 2024–25, [c]Between 2025–26 and 2030–31, [d]Between 2014–15 and 2030–31
Source EFFECT model output

Table 8.25 CO_2 equivalent emissions of passenger and freight transport up to 2030–31 for realistic scenario

Year	CO_2 equivalent emissions (giga gram)							Share (%)			
	Passenger transport			Freight transport			All Total				
	Electricity	Diesel	Total	Electricity	Diesel	Total		Passenger	Freight	Electricity	Diesel
2014–15	5732	3539	9271	8903	4563	13,466	22,737	40.77	59.23	64.37	35.63
2019–20	8139	3940	12,130	11,597	4035	15,633	27,762	43.69	56.31	71.27	28.73
CAGR[a]	7.4	2.17	5.52	5.43	-2.43	3.03	4.07	NA	NA	NA	NA
2024–25	19,467	6349	25,816	31,736	4314	36,050	61,865	41.73	58.27	82.77	17.23
CAGR[b]	18.91	10.01	16.31	22.3	1.35	18.19	17.38	NA	NA	NA	NA
2030–31	55,935	10,132	66,066	95,457	4976	100,433	166,499	39.68	60.32	90.93	9.07
CAGR[c]	19.23	8.1	16.95	20.15	2.41	18.62	17.94	NA	NA	NA	NA
CAGR[d]	15.3	6.79	13.06	15.98	0.54	13.38	13.25	NA	NA	NA	NA

Notes [a]Between 2014–15 and 2019–20, [b]Between 2020–21 and 2024–25, [c]Between 2025–26 and 2030–31, [d]Between 2014–15 and 2030–31
Source EFFECT model output

Table 8.26 CO_2 equivalent emissions of passenger and freight transport up to 2030–31 for pessimistic scenario

	CO2 equivalent emissions (giga gram)							Share (%)			
	Passenger transport			Freight transport			All Total				
Year	Electricity	Diesel	Total	Electricity	Diesel	Total		Passenger	Freight	Electricity	Diesel
2014–15	5732	3539	9271	10,033	5220	15,254	24,524	37.8	62.2	64.28	35.72
2019–20	8057	3913	11,970	22,258	7427	29,685	41,655	28.74	71.26	72.78	27.22
CAGR[a]	7.05	2.03	5.24	17.28	7.31	14.24	11.18	NA	NA	NA	NA
2024–25	18,767	6152	24,919	58,895	7739	66,634	91,553	27.22	72.78	84.83	15.17
CAGR[b]	18.43	9.47	15.79	21.48	0.83	17.55	17.06	NA	NA	NA	NA
2030–31	53,146	9637	62,783	121,206	8208	129,414	192,198	32.67	67.33	90.72	9.28
CAGR[c]	18.94	7.77	16.65	12.78	0.99	11.7	13.16	NA	NA	NA	NA
CAGR[d]	14.93	6.46	12.7	16.85	2.87	14.3	13.73	NA	NA	NA	NA

Notes [a]Between 2014–15 and 2019–20, [b]Between 2020–21 and 2024–25, [c]Between 2025–26 and 2030–31, [d]Between 2014–15 and 2030–31
Source EFFECT model output

a. The total electrical energy requirement of IR could be anywhere between 135,650 MU and 225,206 MU in 2030–31, whereas in 2014–15, the total electrical energy consumption was about 18,248 MU. That is the energy requirement could go up by 7 times to 12 times between 2014–15 and 2030–31.
b. The share of freight transport in the energy requirement was about 55% in 2014–15 and it could go up to 62% in 2030–31, which is mainly due to the huge increase in the freight traffic between 2014–15 and 2030–31. The freight traffic which was 682 billion TKM in 2014–15 is forecast to be between 4634 billion TKM and 7637 billion TKM in 2030–31.

8.9 Emission Profile from Electrical Energy Utilization of Indian Railways

The larger goal of forecasting rail passenger and freight transport and thereby estimating the electrical energy requirement for Indian Railways up to 2030–31 is to find the linkages between the forecasted rail transport and the resultant carbon emissions. In any economic activity, the carbon emission profile varies with energy usage and energy source. Various models are available that examine the linkages between the transport and energy demand in the IR, the sources of energy used in the operations of Indian Railways and the resultant carbon emissions. However, these models are broadly classified into top-down and bottom-up. In the top-down models, macroeconomic theory and econometric techniques are applied to historical data on consumption, prices, incomes and factor costs to model the final demand for goods and services. However, the top-down models address the impact of climate policies on the national and global level and hence have little detail on the energy consuming side of the economy. Bottom-up models focus on engineering gains (which captures technology in the engineering sense) and consider technological options or project specific emission mitigation policies.

8.9.1 Energy Forecasting Framework and Emissions Consensus Tool

Given the nature of this project, the bottom-up models seem to be more suitable given that the context is national, the sector is transport, a subsector of energy consumption and time periods are short term and medium term. Modelling tools were used to forecast greenhouse gas (GHG) emissions from a range of development scenarios. The Energy Forecasting Framework and Emissions Consensus Tool (EFFECT) developed by the World Bank is a bottom-up model to forecast GHG emissions for a range of scenarios in low-carbon development. It focuses on sectors that contribute to and are expected to experience a rapid growth in emissions such as transport. The model was initially developed by the World Bank while working with the Government of India on an analysis of their national energy plan. It is also used for road transport, agriculture, power, industry, household and nonresidential sectors. The EFFECT model is used here to estimate the emissions arising from energy usage in Indian Railways up to 2030–31.

8.9.1.1 The Basic Assumptions for EFFECT Model

a. **The base year**: The base year for the EFFECT model cannot be after 2010. Hence, 2010–11 is taken as the base year.
b. **Categorization of rail transport**: EFFECT divided rail transport into four categories:

(i) Rail Passenger (in turn, divided into three categories):

Electrified Mainline Service (equivalent of electrified non-suburban passenger transport in India)
Suburban Service (Equivalent of electrified suburban rail transport in India)
Nonelectrified Service, Diesel locomotives (equivalent of passenger transport using diesel traction) (both suburban and non-suburban)

(ii) HS Rail Passenger (High-Speed Rail service)
(iii) Rail Metro
(iv) Rail Freight
(v) Electrified Service (equivalent of electrified freight transport in India)
(vi) Nonelectrified Service (equivalent to freight transport using diesel traction)

In India, almost the entire suburban rail service is operated by electric traction and hence Suburban Service is combined with Electrified Mainline Service. Except for the Kolkata Metro, no other metro services are provided by the Indian Railways and hence Kolkata Metro should be considered as part of suburban rail system of IR. High-speed rail services have not yet begun; as indicated earlier, the total energy required per PKM in an HSR will not be 20% more than AC class travel by conventional rail. Hence, the passenger transport associated with HSR was also clubbed with that of the Electrified Mainline Service. Hence, the EFFECT model was used to generate four scenarios,

(i) Rail passenger with electric traction
(ii) Rail freight with electric traction
(iii) Rail passenger with diesel traction
(iv) Rail freight with diesel traction

8.9.1.2 The Input Parameters for EFFECT Model

The various data on IR passenger transport up to 2030–31 such as total passenger transport, share of electric traction, average passenger carriage loading factor, average passenger carriage Annual Mileage, Number of Passenger carriages per train, average mainline locomotives per train, average kms/year travelled per locomotive and their annual change, annual scrappage rate of locomotives and passenger carriages, Energy Efficiency Index of entering locomotives, leaving locomotives and existing locomotives, non-propulsion energy, energy mix of various fuels, etc. were given as input parameters for EFFECT model.

The various data on IR freight transport up to 2030–31 such as total freight transport, share of electric traction, average wagon loading factor, average wagon Annual Mileage, number of wagons per train, average mainline locomotives per train, average km/year travelled per locomotive and their annual change, annual scrappage rate of locomotives and wagons, Energy Efficiency Index of entering locomotives, leaving locomotives and existing locomotives, energy mix of various fuels, etc. were given as input parameters for EFFECT model.

8.9.2 Results of EFFECT Model

The results of the EFFECT model on carbon emissions for the optimistic, realistic and pessimistic scenarios are summarized in Tables 8.20, 8.21 and 8.22.

8.10 Inferences and Conclusions

The key inferences on whether Indian Railways will be able to decarbonize 100% in 2030–31 from the above analysis are summed up as follows:

(i) **Electrified hauling of transport**

With the forecasted growth of electrification of IR tracks, the Indian Railways would be able to haul 80% of passenger transport and 100% freight transport using electric traction in 2030–31. The higher electric traction is prerequisite for IR to decarbonize its operations, as it will give ample scope for IR to source its electricity from non-carbon resources like solar power. If IR wants to achieve 100% decarbonization of its operations, IR should use electric traction for its passenger and freight transport in entirety. For this, a massive electrification of the railway network such as the one proposed by Ministry of Power, whereby 35,000 km of IR network would be electrified within 3 years (Economic Times Mar 03 2016), will have to be planned else it will not be possible to achieve 100% electric hauling of passenger and freight transport by 2030–31.

(ii) **In-vehicle energy consumption in passenger transport**

With increasing demand for AC travel, about 45–59% of passenger transport in 2030–31 would be by AC classes. As the in-vehicle energy required in an AC class is about 70% of the hauling energy required for hauling an AC passenger car, the overall energy consumption per unit of passenger transport would increase significantly up to 2030–31. Rooftop solar panels may partially meet the in-vehicle consumption of AC coaches, at least during the daytime.

(iii) **Electrical energy required for rail transport and the solar power capacity to be created**

The usual benchmark for energy generated from a 1 MW solar power plant is 1.5 million units (Zolt energy n.d). To generate all the electrical energy required by the railways, solar capacity to the tune of 90,000–150,000 MW will have to be installed by 2030–31, generating 136–225 billion units. By switching to solar power, Indian Railways would generate carbon emissions only from about 10% of its total operations.

References

Andersson E, Lukaszewicz P (2006) Energy consumption and related air pollution for Scandinavian electric passenger trains, Report KTH/AVE 2006:46 Stockholm, 2006. Retrieved from https://www.ave.kth.se/polopoly_fs/1.179879!/Menu/general/column-content/attachment/Energy_060925_full_pdf.pdf

Bhutada S (2016, April) Mumbai gets its first-ever AC local train. The Indian Express. Retrieved from http://indianexpress.com/article/cities/mumbai/mumbai-gets-its-first-ever-ac-local/

Biswas S (2016, April 23) All aboard Talgo: 10 things to know about the high speed Talgo trains coming to India. India today.in. Retrieved from http://indiatoday.intoday.in/story/talgo-trains-all-you-need-to-know-high-speed-spanish-trains-come-to-india/1/649659.html

Economic Times (2016, March 03) Government mulls electrification of 35,000 km rail line in 3 years. Retrieved from http://economictimes.indiatimes.com/industry/transportation/railways/government-mulls-electrification-of-35000-km-rail-line-in-3-years/articleshow/51244115.cms

Gangwar M, Sharma SM (2014) Evaluating choice of traction option for a sustainable Indian Railways. Transp Res Part D 33(2014):135–145

Indian Railways Green Energy Initiatives (2015) Future initiative to improve energy utilization in electric traction. Retrieved from http://www.irgreenri.gov.in/tile_led.html

Ministry of Railways (n.d.c) Financial statements & operating statistics 2004–05. Annual Report & Accounts 2004–05. Retrieved from http://www.indianrailways.gov.in/railwayboard/uploads/directorate/stat_econ/2004_05/AR_04_05/financial_statements_Operating_Statistics.pdf

Ministry of Railways (n.d.j) Financial statements & operating statistics. Indian Railways Annual Report & Accounts_2014–15. Ministry of Railways (Railway Board). Government of India. Retrieved from http://www.indianrailways.gov.in/railwayboard/uploads/directorate/stat_econ/IRSP_2014-15/IR_Annual_Report26Accounts_2014-15/Financial_Statements.pdf

MoEF (2010) Annual Report 2009–2010. Ministry of Environment and Forest. Government of India, Delhi

MoR (2009) Indian Railways Vision 2020. Railway Board, Ministry of Railways, Government of India. Retrieved from http://www.prsindia.org/uploads/media/RailwaysVisionDocument2020.pdf

MoR (2015) Indian Railways Lifeline of the nation. A White Paper, February 2015. Retrieved from http://www.indianrailways.gov.in/railwayboard/uploads/directorate/finance_budget/Budget_2015-16/White_Paper-_English.pdf

MoR (n.d.a) Passenger Business - Indian Railways up to 2010–11. Ministry of Railways (Railway Board). Government of India. Retrieved from

MoR (n.d.b) Traction—Indian Railways up to 2010–11. Ministry of Railways (Railway Board). Government of India. Retrieved from http://indianrailways.gov.in/railwayboard/uploads/directorate/stat_econ/yearbook10-11/Traction.pdf

Network Rail (n.d) Comparing environmental impact of conventional and high speed rail. www.networkrail.co.uk/5878_Comparingenvironmentalimpactofconventionalandhig

NTDPC (2014) India Transport Report—Moving India to 2032. Volume II Main Report. National Transport Development Policy Committee, Jan 31, 2014. Retrieved from http://planningcommission.nic.in/reports/genrep/NTDPC_Vol_02.pdf

OECD/IEA (2015) Energy and Climate Change, World Energy Outlook Special Report. Retrieved from https://www.iea.org/publications/freepublications/publication/WEO2015SpecialReportonEnergyandClimateChange.pdf

Office of the Chief Economist (2015) Coal in India 2015. Office of the Chief Economist. Department of Industry and Science. Australian Government. June 2015. Retrieved from https://www.industry.gov.au/Office-of-the-Chief-Economist/Publications/Documents/Coal-in-India.pdf

Parikh et al (2014) The final report of the expert group on low carbon strategies for inclusive growth. Planning Commission. Government of India. Retrieved from http://planningcommission.nic.in/reports/genrep/rep_carbon2005.pdf

PIB (2015, August 7) Progress in dedicated freight corridor. Press Information Bureau. Ministry of Railways. Government of India. Aug 07 2016. Retrieved from http://pib.nic.in/newsite/PrintRelease.aspx?relid=124598

PIB (2016, November 3) Minister of Railways inaugurates International Conference on Decarbonisation of Indian Railways—Mission Electrification. Press Information Bureau. Aug 07, 2016. Government of India. Ministry of Railways. Retrieved from http://pib.nic.in/newsite/PrintRelease.aspx?relid=153230

Ramakrishnan TS (2015, July 17) Special Trains, Special Fares. The Pioneer. Retrieved from http://www.dailypioneer.com/columnists/oped/special-trains-special-fares.html

Statistics Times (2016, February 18). GDP growth of India. Statistics Times collated from Planning Commission, Ministry of Statistics and Programme Implementation, Economic Survey of India 2014–15 and International Monetary Fund. Retrieved from http://statisticstimes.com/economy/gdp-growth-of-india.php

UNDP (2015) Improving Energy Efficiency in Indian Railways, UNDP. Retrieved from http://www.in.undp.org/content/dam/india/docs/improving_energy_efficiency_in_the_indian_railway_system_factsheet_project.pdf

World Bank (n.d) CO_2 emissions from transport (% of total fuel combustion) collated from IEA Statistics between 1960 and 2014. Retrieved from http://data.worldbank.org/indicator/EN.CO2.TRAN.ZS

Zolt energy (n.d) Frequently Asked Questions on Solar Power. Retrieved from http://www.zoltenergy.co/

Ministry of Railways (Railway Board) (2015a) Explanatory Memorandum on the Railway Budget for 2015–16

Ministry of Railways (Railway Board) (2015b) Outcome and Performance Budget of railways for 2015–16

Chapter 9
Pay Less for More: Energy Efficiency Approach to Municipal Water Supply in Indian Cities

Indro Ray

Abstract As more people start living and moving to urban areas, and with improving lifestyles and economic prosperity (urban monthly per capita expenditure of Rs. 2630 was 84% higher than rural expenditure in 2011–12), the demand for finite resources such as land, water, and fossil fuels will be higher than ever. India is one of the most water-scarce countries in the world. Indian municipalities are undercapitalized and their high energy costs are neither feasible nor sustainable. To bring down the energy cost, the water sector in urban areas needs interventions in both demand and supply sides, and at individual and institutional levels. This paper focuses on various energy efficiency measures to reduce energy costs for water supply at the municipal level in India and calculates the magnitude of these cost savings and reduction in emissions for 10 of the 53 large Indian cities with population more than a million in 2011. Showcasing these energy cost savings and GHG reduction findings, this research argues for widespread adoption of energy efficiency measures as it brings multiple benefits.

9.1 Introduction

In 1990, one out of four Indians lived in its cities. According to the latest census of 2011, this ratio has increased to one out of three. In the last decade India's urban population grew twice as fast as its total population. At present, urban areas host more than 370 million Indians. Based on this scale and urban growth trajectory, the UN urbanization prospect projected urban population in India by 2050 will reach close to 700 million. As more people start living and moving to urban areas, and

I. Ray (✉)
Rio Tinto, New Delhi, India
e-mail: rayindro.asu@gmail.com

R. Kathuria et al. (eds.), *Low Carbon Pathways for Growth in India*, India Studies in Business and Economics, https://doi.org/10.1007/978-981-13-0905-2_9

with improving lifestyles and economic prosperity (urban monthly per capita expenditure of Rs. 2630 was 84% higher than rural expenditure in 2011–12),[1] the demand for finite resources such as land, water, fossil fuels will be higher than ever. Among these, water being one of the critical lifelines for humans needs immediate attention. In a country like India, which is one of the most water-scarce countries in the world, high water demand, increasing pressure on existing sources, and unmanaged usage have resulted in rapid depletion of water tables and drying of rivers and canals. Given low water availability, discharge of untreated wastewater into freshwater bodies around large settlements is also a big concern. The cumulative effect of all these issues has resulted in decrease in per capita availability of water in India by more than 22% (ESCAP 2009)[2] in the last decade. By 2050, per capita water availability will come down to 60% of 2001 levels.[3] The urban local governments that are mandated to supply clean water to its citizens face serious challenges with the quantity and quality of available water. They also struggle with high costs of delivery. When it comes to the cost of water supply, the single largest expense is the cost of energy needed to supply water (from source to tap). This cost can be as high as 60% of the total electricity bill for the whole municipality (Rao and Sharma 2013),[4] but it usually ranges between 30 and 50%, which is also very high (Planning Commission 2011).[5] This trend is common in most developing countries where the ratio of cost of energy for water supply to the municipality's energy budget is very high.

Indian municipalities are undercapitalized and their high energy costs are neither feasible nor sustainable. To bring down the energy cost, the water sector in urban areas needs interventions in both demand and supply sides, and at individual and institutional levels. These interventions may take the form of water conservation, recycle, and reuse practices, energy efficient technologies, better decision-making in procuring equipment, financing from government and international agencies, and active public participation. With the objective of reducing energy costs for water supply at the municipal level, this paper focuses on various energy efficiency measures. Combination of many of these measures can bring down the annual total energy cost in the range of 25–40%.[6] Such cost saving will strengthen the financial health of local governments and as a co-benefit will provide significant environmental benefits in the form of less greenhouse gas (GHG) emissions. The magnitude of these cost savings and reduction in emissions are calculated for 10 of the 53 large Indian cities with population more than a million in 2011.

[1]National Sample Survey Office, Ministry of Statistics and Programme Implementation, Government of India (2013), http://mospi.nic.in/Mospi_New/upload/press-release-68th-HCE.pdf.

[2]ESCAP (2009).

[3]Water Sector in India: Overview and focus areas for the future (2010). KPMG, India. https://www.kpmg.de/docs/Water_sector_in_India.pdf.

[4]Rao and Sharma (2013).

[5]Planning Commission (2011).

[6]Energy Sector Management Assistance Program (2012).

9.2 Scale of the Urban Water Supply Problem

Recent data on India's urbanization, distribution of urban population, and water stress show that the future of water in urban India is bleak unless appropriate and timely measures are taken. The distinct type of urbanization in India is marked by high influx of migrants from rural to urban areas. This has resulted in a distributed model of urbanization with diverse range of large and small size cities spread across the country.[7] The current urban population of 377 million people reside in 7933 cities and towns.[8] It is worth noting that more than 2/3rd of urban population is concentrated in 180 districts (out of total 641 districts). These regions also have urban population greater than the national average of 32.16% in 2011 reinstating the fact that urban population are concentrated in few urban agglomerations. Figure 9.1 shows the distribution of urban population across all the districts in India, which clearly highlights the uneven distribution of urban population.

The cities and urban agglomerations that are sites of rapid population growth have lacked behind in providing services and infrastructure to all its residents. According to India's 12th Five Year Plan (2012–2017), an average of 8.5–9% of India's GDP will be required annually for infrastructure development in India over the 5-year period. This gap in demand and supply has created multiple challenges. One of the most critical challenges is to provide sufficient and clean water with limited resources and a weak and inadequate existing distribution system. In 2011, only 62% of urban households had access to treated tap water and more than one-third still depended on groundwater. Figures 9.2 and 9.3 show percentage of urban households with access to drinking water through tap and groundwater in different districts in 2011. These figures highlight the clear developmental divide between the eastern and western parts of India. Urban households in most of the southern states of Tamil Nadu, Karnataka, Andhra Pradesh, and western states of Maharashtra, Gujarat, and Rajasthan have piped water connection. In a stark contrast, central and eastern states of Uttar Pradesh, Madhya Pradesh, Chhattisgarh, Jharkhand, Bihar, West Bengal, and Assam have very few households with tap water access. The lack of piped water is compensated through use of groundwater. This case particularly stands out for the state of Bihar. Overall, the state of urban water network in India is poor. This poor condition is a result of low investment in urban infrastructure, which stands at $17 per person per year as compared to $100 suggested by most benchmarks.[9] The case of poor access to water also comes through the quantitative analysis of urban analysis.

While analyzing the degree of association between urbanized areas and water access through tap, it is found that though the correlation is positive (0.38), it is not

[7]Planning Commission. (2012).

[8]Census of India (2011).

[9]Planning Commission. (2012).

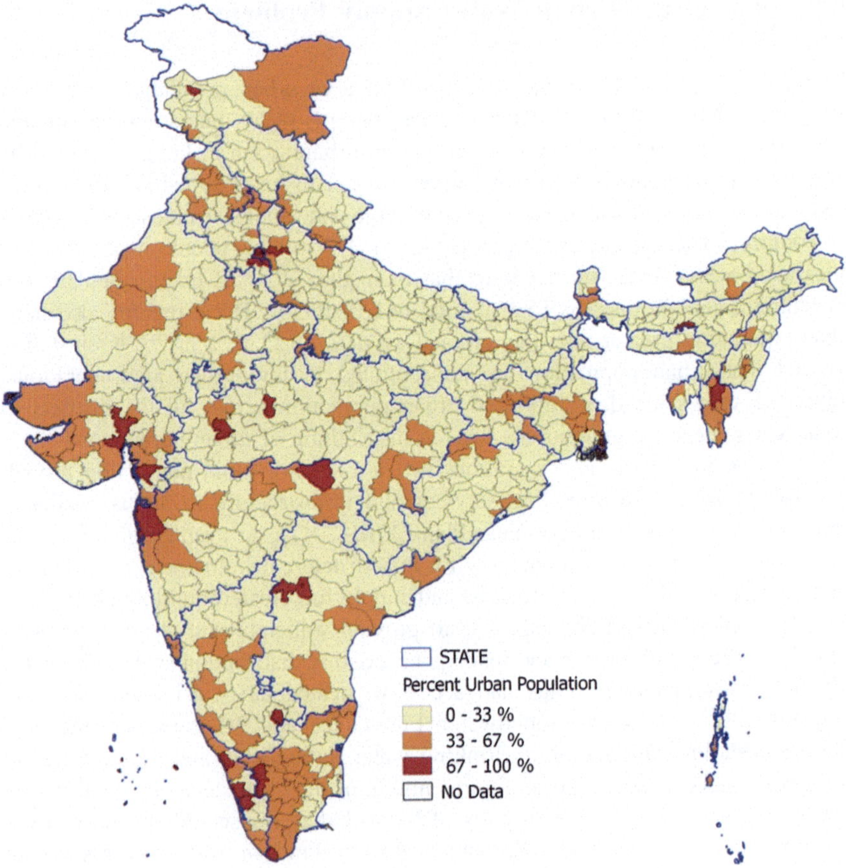

Fig. 9.1 District level urbanization (2011). *Source* Census of India (2011); Author

very high. This indicates that a large percentage of urban households still depend on groundwater or any other source.

Households in urban areas in India not only suffer from poor access to water but urban regions also have low freshwater availability. Figure 9.4 shows baseline water stress in India.[10] Water stress is calculated using 2010 data for total annual water withdrawal and total water supply. This measure indicates a relative water demand. The World Resources Institute's Aqueduct Program claims that 54% of the country faces high to extremely high levels of water stress.[11] Juxtaposing urban

[10]Baseline water stress measures total annual water withdrawals (municipal, industrial, and agricultural) expressed as a percent of the total annual available flow. Higher values indicate more competition among users. Arid areas with low water use are shown in gray, but scored as high stress when calculating aggregated scores.

[11]Shiao et al. (2015).

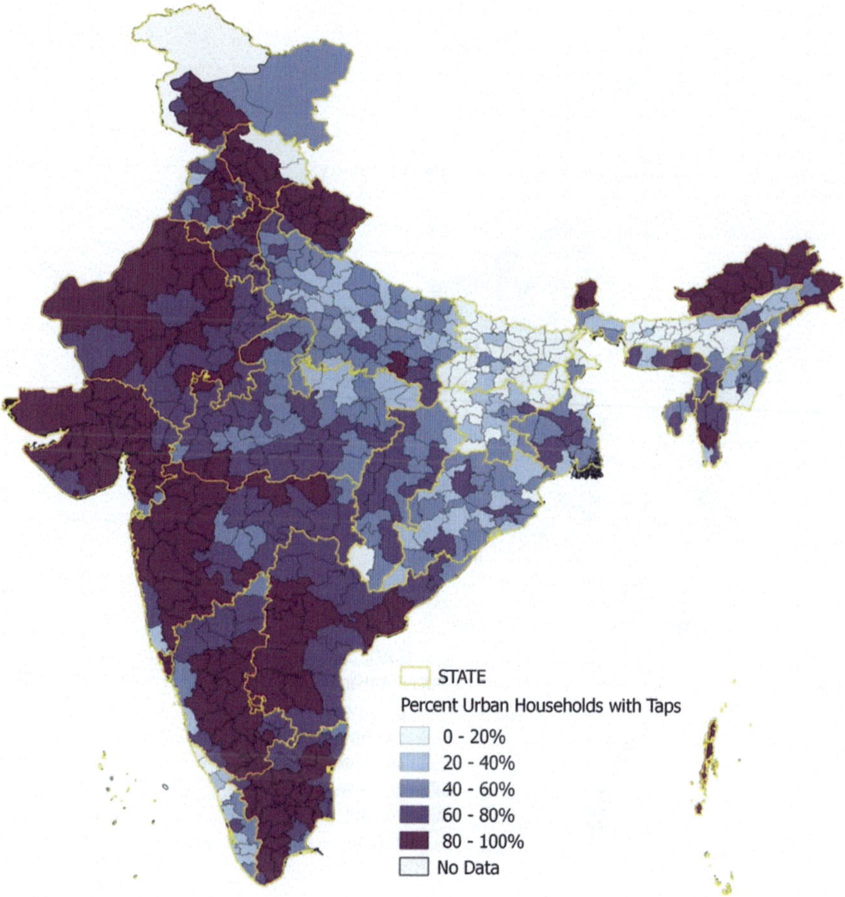

Fig. 9.2 Access to drinking water through taps in urban households (2011). *Source* Census of India (2011); Author

areas with high dependence on groundwater or non-surface water (as seen in previous figures) and high water stress regions reveal an alarming picture. Most of the urban districts are located in high-stress regions.

Given this existing condition and certainty of high demand for water in the future, it seems clear that local governments will face major hurdles on account of low water availability. Other factors such as high pollution levels, high temperatures, declining groundwater levels, and irregular rainfall patterns will make the situation worse. In addition, urban local bodies are cash strapped and are unable to meet the high energy costs associated with water delivery. At present, total water supplied in Indian cities stand at 56,000 million liters per day (MLD), which translates to 20,440 billion liters in a year.[12] To supply such large volume of water,

[12]Excreta Matters-Volume I (2012). Center for Science and Environment, New Delhi.

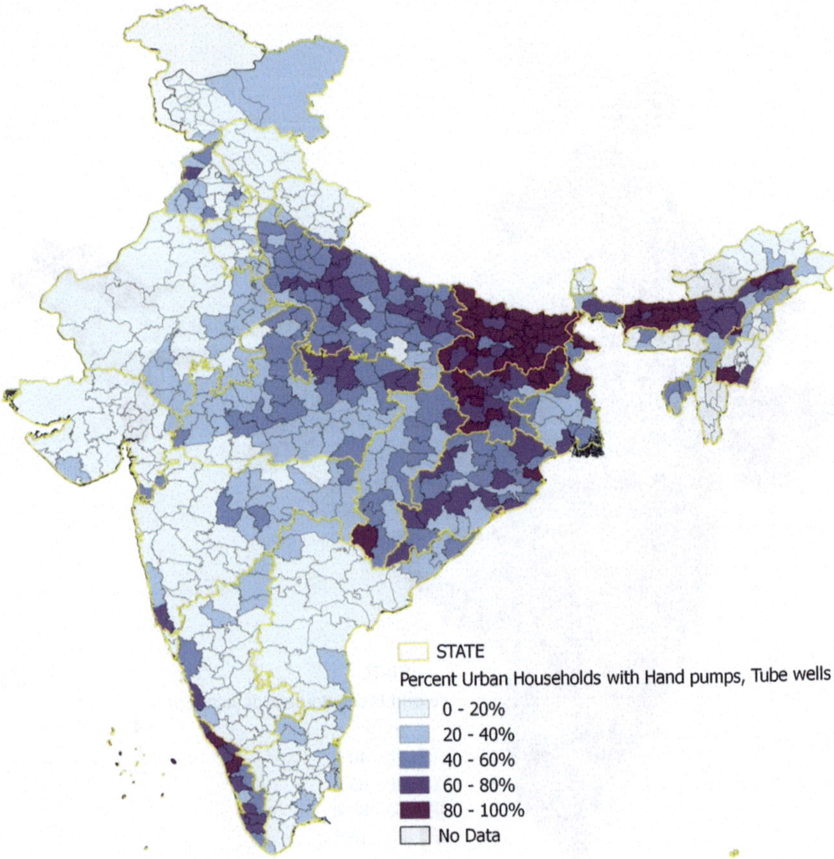

Fig. 9.3 Access to drinking water from groundwater for urban households (2011). *Source* Census of India (2011); Author

public water works in cities cumulatively consume around 17,000 MU (million units) of electricity annually and it is projected to be up to 37,000 MU by 2021–22.[13] With water demand projected to increase threefold by 2030, adopting energy efficiency measures to save energy and reduce the energy costs of delivering water becomes essential. Before discussing these measures, it will be worthwhile to look at energy consumption patterns in a water supply network at the city level so that one can understand the scope and scale of cost savings.

[13]18th Electric Power Survey of India, Central Electricity Authority. Ministry of Power, Government of India.

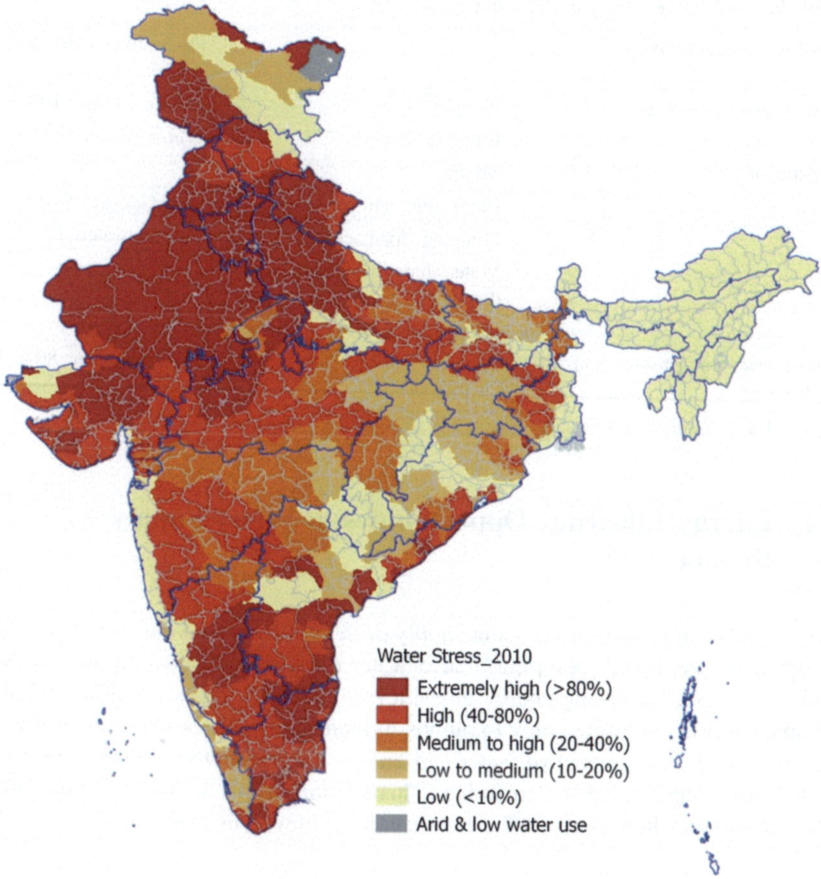

Fig. 9.4 Baseline water stress (annual withdrawal as percentage of annual availability) (2010). *Source* Aqueduct: measuring and mapping water risks World Resources Institute (2010); Author

9.3 Energy Consumption in Municipal Water Supply System

Electricity (or energy) is essential for most of the functioning of an urban water supply system. It is used to pump raw untreated water from different sources, water treatment, storing water in reservoirs, and finally distributing treated water to end users. The table below breaks down the energy consumption pattern for different stages of water supply by source (Table 9.1). Energy usage varies for different sources of water, i.e., surface and groundwater. Among all the steps in the water use cycle, as one can see in the table below, pumping of water is most important and it is the most energy-intensive part.

Table 9.1 Indicative energy use of municipality water

Energy using activity		Indicative energy use share
Raw water extraction	Pumping	Surface water: 10%
	Building services	Groundwater: 30%
Treatment	Mixing	
	Other treatment processes	Surface water: 10%
	Pumping (for backwash, etc.)	Groundwater: 1%
	Water sludge processing and disposal	
	Building services	
Clean water transmission and distribution	Pumping	Surface water: 80%
		Groundwater: 69%

Source The Energy Sector Management Assistance Program (ESMAP), World Bank.

9.4 Energy Efficiency Opportunities in Water Supply System

Pumps and motors used in the water supply cycle usually run all day long every day of the week. As noted, pumping treated water consumes 70–80% of all the electricity used. So, increasing energy efficiency definitely includes installing efficient pumps, but it is not limited only to pumps. It includes measures like using software for automated management of water and energy use (SCADA), reducing physical leakage of water, and water metering. This section briefly looks into these areas while discussing how energy efficiency can be achieved in cities.

9.4.1 Supervisory Control and Data Acquisition (SCADA)

Before implementing any steps aimed at increasing energy efficiency, it is necessary to understand where, how much, and when energy is being consumed in the water supply system. Energy audits can establish baseline energy usage and the collected data can be fed into a SCADA system (a computer-based system used to monitor and control industrial, infrastructure, or facility-based processes) which will show where and how much energy is being used. City managers can use this information to identify energy-intensive components and ways to bring about improvements in energy use. The U.S based Electric Power Research Institute (EPRI) has estimated that 10–20% of energy can be saved by optimizing waterworks system processes using SCADA.

9.4.2 Pumps, Pipes and Other Equipments

In the pursuit to save energy cost, one of the first things municipalities can do is upgrade their equipments. Pumps and blowers are most extensively used to extract raw water and distribute treated water into the supply network. Most of the times pumps and often other electrical and mechanical equipments are used beyond their expected life, resulting in energy inefficiency. In such cases, pumps need be replaced with more efficient and upgraded models. Buying equipment of the right size is also important. Most often cities buy and lay down oversized pipes and install large capacity pumps to accommodate future load and save on buying larger pumps later on. But studies by the Centre for Science and Environment show that energy efficient pumps rather than higher capacity pump are more cost-effective in the long run.

Energy efficiency can also be achieved through installing and running two or more pumps in parallel. This is also one of the efficiency methods recommended by the Bureau of Energy Efficiency in India. When water is needed to be pumped up to a height, two or more pumps can run in parallel. The variation in flow achieved by switching on and off addition pumps can meet municipal water demand and flow control. In this process if care is taken while running pumps to keep their operation within prescribed functional limits as set by the manufacturer, it can be energy efficient while meeting the demand.[14] Besides moving onto efficient and additional pumps, cities can significantly reduce the cost of pumping by shifting the use of electricity away from peak demand times to off-peak hours.

9.4.3 Water Metering

In developing counties, most cities lack any kind of water metering. Large quantities of water are thus unaccounted for leading to water wastage and loss of revenue. But if water meters are installed they can measure volumes of water used and account for it. Water meters thus not only help in managing water consumption but also bring efficiency and cost cutting. To be even more efficient and beneficial, meters should be automated, accurate, and should be repaired in a timely manner. But as Indian cities severely lack water metering, they need to take on this activity aggressively. Though this process is capital intensive, it needs to be done to reap multiple benefits and for this, central and state governments can provide subsidy to cities. A 2009 study by The Energy and Resources Institute (TERI) for Indian cities claims that metering and energy cost accounting have helped many municipalities cut their energy consumption by almost 10%.[15]

---

[14]Bureau of Energy Efficiency (2005).

[15]The Energy and Resources Institute (2009).

9.4.4 *Leakage Reduction*

Besides the issue of lack of water metering, Indian cities are marred by old infrastructure, illegal tapping of pipelines, pilferages, and other problems. Water leakage due to these problems accounts for almost 40% of the total volume of water supplied in most Indian cities.[16] According to best global practices, this percentage should be below 10.

Since cities cannot claim revenue from water lost through leakage, it is termed as nonrevenue water (NRW). Loss of such large quantities (up to 40%) of usable water also hampers municipal service delivery and places an additional burden through energy consumption costs. Thus, tackling water leakages by replacing old pipelines, and quickly detecting and addressing leakages through technologies (such as GIS and RADAR) can provide multiple benefits. Through this single measure, it is possible to increase water supply while utilizing existing water resources and save on energy cost.

9.5 Energy Consumption and Cost Savings, and CO_2 Reduction

In 2013, The Ministry of Power, Government of India came out with its 18th edition of the Electric Power Survey of India. The survey provides data on electrical energy consumption, energy requirements, and peak loads for India and its 13 large cities. Under the energy consumption category, data are provided for domestic, commercial, public lighting, public waterworks, irrigation, heavy and light industries, and railway traction.[17] To compute energy efficiency in urban water supply, energy consumption in public water works for the year 2010–11 is considered for 10 of the 13 cities (see Table 9.2). Three metro cities (Kolkata, Chennai, and Mumbai) are not considered because of some missing data points and to avoid skewness in results. Delhi was not part of the survey.

The 2011 average state power tariffs for public works come from the Power and Energy Division of the Planning Commission. In the absence of power tariffs for the public sector, rates for "railway traction" are used to calculate the total energy expenditure on public water works for the cities. Indian Railways being a public sector department, it is assumed that its power tariffs will be close to those paid by the cities to buy electricity. All the calculations for energy savings and subsequent cost savings are based on these tariff rates. The reduction in CO_2 emissions (due to less use of energy) is based on the weighted average emission factor of 0.79 ton

---

[16] Excreta Matters—Volume I (2012). Centre for Science and Environment, New Delhi.

[17] 18th Electric Power Survey of India, Central Electricity Authority. Ministry of Power, Government of India.

Table 9.2 Energy, cost savings and CO_2 reduction in large Indian cities

S. No.	Cities	2011 Population	Power consumed by Public Water Works (in MWh)	Average Power Tariff (in Rs./Kwh)	Annual Energy Cost for Public Water Works (in Rs. Crores)	25% less Energy Consumed (in MWh)	40% less Energy Consumed (in MWh)	25% Energy Cost Saving (in Rs. Crores)	40% Energy Cost Saving (in Rs. Crores)	CO2 reduced with 25% saving (in ton/MWh)	CO2 reduced with 40% saving (in ton/MWh)
1	Indore	1,960,631	12	5.00	6.00	3.00	4.80	1.50	2.40	2.37	3.79
2	Nagpur	2,405,421	44	6.59	28.98	11.00	17.60	7.25	11.59	8.69	13.90
3	Kanpur	2,767,031	50	4.12	20.62	12.50	20.00	5.15	8.25	9.88	15.80
4	Lucknow	2,815,601	129	4.12	53.20	32.25	51.60	13.30	21.28	25.48	40.76
5	Jaipur	3,073,350	126	4.04	50.85	31.50	50.40	12.71	20.34	24.89	39.82
6	Pune	3,115,431	231	6.59	152.16	57.75	92.40	38.04	60.86	45.62	73.00
7	Surat	4,462,002	112	5.58	62.45	28.00	44.80	15.61	24.98	22.12	35.39
8	Ahmadabad	5,570,585	108	5.58	60.22	27.00	43.20	15.05	24.09	21.33	34.13
9	Hyderabad	6,809,970	90	4.74	42.69	22.50	36.00	10.67	17.08	17.78	28.44
10	Bangalore	8,425,970	362	4.29	155.20	90.50	144.80	38.80	62.08	71.50	114.39
Average		4,140,599	126	5.06	63.24	31.60	50.56	15.81	25.29	24.96	39.94

Source Ministry of Power; Planning Commission; Author

CO_2/MWh for India for 2010–11. The figure for emission factor is provided by the Ministry of Power.[18]

Based on the discussion in the previous sections, existing literature and cases, it is claimed that implementation of energy efficiency measures in municipal water supply can bring down energy consumption by 25–40%. Considering these numbers as the range for reducing energy consumption, Table 9.2 provides annual energy savings and CO_2 reduction for each of the large 10 cities. Based on the 2010–11 energy consumption in respective municipal water supply and power tariff figures, cities on average spent Rs. 156 per person per year for energy. Among these cities, Indore pays the lowest amount (Rs. 31) while Pune pays the highest amount (Rs. 488) per capita for energy in a year. Under the existing conditions, energy efficiency measures if implemented will save Rs. 38–Rs. 61 per person. This translates into a reduction of energy costs of Rs. 15–25 crores per year per city. Besides the monetary benefits, such savings will help India achieve a low carbon pathway to which it has pledged voluntary commitment. The annual average reduction in CO_2 emissions will vary with the extent of energy saved. Consuming an average of 25% less energy will reduce CO_2 emissions by 6.03 g/MWh per person in a year. With 40% savings of energy, this figure stands at 9.65 g/MWh per person.

9.6 Case for Energy Efficiency in Municipal Water Supply Systems

The cities considered in the energy saving and CO_2 reduction calculations are large cities with an average population of more than 4 million. While it is known that small and medium size urban areas hold a large percentage of urban population as well. So, in this case, generalizing these results to the national level may not be fruitful given the small sample of large sized cities. But they indeed showcase the scope of benefits that can be achieved in small and medium cities through improvements in energy efficiency in the public water distribution network.

The latest census data show there are 468 towns in India that have a population greater than 1 lakh including 53 million-plus cities. Based on the Census of India definition for urban areas and the fact that there are 4041 statutory towns of all sizes, it is calculated that there are 3573 statutory towns with population between 5000 and 1 lakh. There are close to 155 million people living in these relatively small statutory towns. These towns are notified under law by their respective state governments and have local bodies like municipal corporations, and municipalities. By law, these local bodies are responsible for providing essential services including water to residents and businesses. The current situation in India is that the large cities have their own water departments while the census towns get water from their

[18]Central Electricity Authority, Ministry of Power, Government of India.

respective state's water departments through pipes and community taps. This leaves the small and medium sized statutory towns with inadequate and inefficient infrastructure to provide water. They also lack enough revenue and funds to invest and augment their water supply networks. Even in cities that have piped water networks, the coverage is partial, leading to inequality in access. In such cases, water demand is mostly met by pumping groundwater through tube wells, bore wells, or hand pumps by individual households. Patel and Krishnan (2008) found that urban areas with populations between 10,000 and 1 lakh depend mostly on groundwater (which is more energy intensive as noted in Table 9.1) while cities with more than a million people depend the least on groundwater.[19] The findings in this paper and the existing literature thus make a good case for expanding piped water network in medium size cities while incorporating different energy efficiency measures.

With time as small and medium size cities increase in size, the demand for better service delivery will grow. If local governments intend to meet this high demand by pumping groundwater or transporting water over long distances, it will be a resource-intensive endeavor (both in water and energy usage terms). To avoid or minimize such costs, reduce dependency on groundwater, curtain energy inefficiency and inequality in water access, city governments should adopt measures that are environmentally friendly. To save on costs, they should strive for choices that are economical in the medium run (1–3 years) rather than going for immediate cost savings through low-priced but energy-intensive equipments. Due to the magnitude of this problem of high energy cost, it has received attention from both private and public sector. The private sector is coming up with low energy use water pumps while public agencies like Bureau of Energy Efficiency have come up with energy efficiency ratings for pumps. While augmenting water supply network, cities also need to invest in programs such as installing water meters, timely addressing water leakages, and taking appropriate administrative steps.

9.7 Conclusions

In the next few decades, the scope of urban growth is massive in the large number of small and medium size statutory towns in India. To catch up to this growth, the local municipal governments not only have to augment their existing infrastructure but invest heavily in providing adequate level of services. One such service that needs urgent attention is water supply as the existing provisions are insufficient and it costs the local government large sums of money. The cost of energy for water supply can be up to 50% of the total annual electricity budget for a city.

In a water stress country like India, to meet the growing demand of water in its cities, local governments must adopt energy efficiency measures. These include

[19]Patel and Krishnan (2008).

buying marginally expensive but energy efficient pumps and other equipments, adopting automated water accounting SCADA system, installing water meters, and timely addressing water pilferages and leakages. Adopting some or all these measures can cut the energy cost for water supply between 25 and 40%. This translates into an approximate annual average cost saving of Rs. 50 per person in cities with added benefit of per person 7.5 g/MWh CO_2 emission reduction. Showcasing these energy cost savings and GHG reduction findings, this research argues for widespread adoption of energy efficiency measures as it brings multiple benefits. Energy efficiency brings down service delivery cost in the long run while resulting in cleaner and better environment. For these reasons alone, selling the idea of sustainable practices such as energy efficiency to citizens, administrators, and policymakers should not be very difficult.

Acknowledgments This work was conducted in his time at the Indian Council for Research on International Economic Relations (ICRIER) and does not represent Rio Tinto's views.

References

Bureau of Energy Efficiency (2005) Guidebook: electrical efficiency in electrical utilities

Census of India (2011) Primary Census Abstract. Accessed from http://www.censusindia.gov.in/2011census/PCA/PCA_Highlights/pca_highlights_file/India/5Figures_at_glance.pdf

Energy Sector Management Assistance Program (2012) The international bank for reconstruction and development. The World Bank Group

ESCAP (2009) Statistical Year book for Asia and the Pacific

Patel A, Krishnan S (2008) Groundwater situation in urban India: overview, opportunities and challenges. In: ICFAI (ed), Water supply and sanitation: essentials of urban economic growth, ICFAI Press

Planning Commission (2011) Report of the working group on urban and industrial water supply and sanitation for the twelfth five-year-plan (2012–2017). Accessed 21 Mar 2015 from http://planningcommission.nic.in/aboutus/committee/wrkgrp12/wr/wg_indu_sani.pdf

Planning Commission (2012) Approach to the 12th plan: the challenges of urbanization in India

Rao RG, Sharma VK (2013) Energy cost savings in municipal water pumping systems-need for web interactive tool. Int J Sci Environ Technol 2(5):920–929

Shiao T Maddocks A, Carson C, Loizeaux E (2015) 3 Maps explain India's Growing water risks. World Resources Institute. http://www.wri.org/blog/2015/02/3-maps-explain-india%E28099s-growing-water-risks

The Energy and Resources Institute (2009) Best practices guide for energy efficiency in municipal water pumping

Chapter 10
Sunny Side up: India's Journey to Energy Security

Indro Ray

Abstract Environmental degradation coupled with the threat of climate change cast a dark cloud over India's future economic growth. Many of the large Indian cities have highest level of air pollution concentration in the world. There is also a large section of poor population that still depends on biofuel for their heating and energy purpose at the household level, most farmers practice burning of crop residue at the end of the harvest season, and much of India's energy is till generated from coal-based thermal power plants. India's conventional sources of power generation are depleting fast and a situation leading up to an energy crisis is probable. To avoid such situation while mitigating the climate change risks, in the last decade, Indian government has focused on generating power using renewable sources (both grid-connected and off-grid) and to complement its efforts, it has devoted considerable amount of resources and political mandate towards this goal. This research tries to document the role of renewable energy, especially solar energy sector's history, plot its recent trends, and understand the future challenges.

10.1 Introduction

Since India's economic liberalization in early 1990s, its economy has grown rapidly and in most aspects this growth has been inclusive. The one area where some work still needs to be done is environmental protection. Environmental degradation coupled with the threat of climate change cast a dark cloud over India's future economic growth. Consumption of fossil fuel has increased over the years leading to high level of pollution with costly negative externalities. Many of the large Indian cities have highest level of air pollution concentration in the world. There is also a large section of poor population that still depends on biofuel for their heating and energy purpose at the household level, most farmers practice burning of crop residue at the end of the harvest season, and much of India's energy is till generated

I. Ray (✉)
Rio Tinto, New Delhi, India
e-mail: rayindro.asu@gmail.com

R. Kathuria et al. (eds.), *Low Carbon Pathways for Growth in India*, India Studies in Business and Economics, https://doi.org/10.1007/978-981-13-0905-2_10

from coal-based thermal power plants. These activities add to the worsening of the environment with high health cost. India is also facing climate change impacts as in its recent history it has seen some deviation from its normal climatic patterns. There has been an increase in its surface temperature, variation in its monsoon patterns has been observed, the sea level is rising by 1–2 mm per year, extreme weather events have become more frequent, and the water flow in its perennial rivers has become more erratic. On top of these factors, India's conventional sources of power generation are depleting fast and a situation leading up to an energy crisis is probable. To avoid such situation while mitigating the climate change risks, in the last decade, Indian government has focused on generating power using renewable sources (both grid-connected and off-grid) and to complement its efforts, it has devoted considerable amount of resources and political mandate towards this goal. In terms of policy and programs, in 2008, the National Action Plan for Climate Change (NAPCC) was launched that officially kick started India's renewable energy and energy efficiency goals. Since then, it also brought out the Intended Nationally Determined Contributions (INDCs) in which it outlined climate actions India intended to take. At the global platform, it is actively engaged in negotiations in the UN Framework Convention on Climate Change (UNFCCC) and most recently committing to its INDC goals, it has ratified the Paris Climate Agreement in COP 21. With such progressive actions and rapid change in India's renewable energy, especially solar energy scene as the background, it is worth to document its history, plot its recent trends, and understand the future challenges.

10.2 Growth of Renewable Energy Share in India's Energy Supply Mix

In the last decade and half (over three 5-year plans), the electricity generation in India has seen much growth. The CAGR during the 10th plan was 5.16% and in the 12th plan (2012–17), it increased to 6%. Such growth has resulted in a situation where electricity supply potential is greater than the economic demand. This is the first time such situation has occurred in Indian history. Much of the success factor lies with the deep penetration of renewable energy (mainly solar and wind energy) in the energy mix. This has come through mainly due to the focus on climate change, energy security concerns, long-term sustainability, enhanced electricity access, and reliability of power.[1] The transformative role of renewable energy in India can be seen through its growth in the last 30 years (see the graph below). Though coal-based thermal electricity generation has hovered around 60%, the share of renewable energy has increased from 0% in 1985 to 15.6% in 2017, which translated to 50 GW. It is the first time that share of electricity generated from hydropower is lower than renewable sources (Fig. 10.1).

[1]TERI (2017). Transitions in Indian Electricity Sector: 2017–2030.

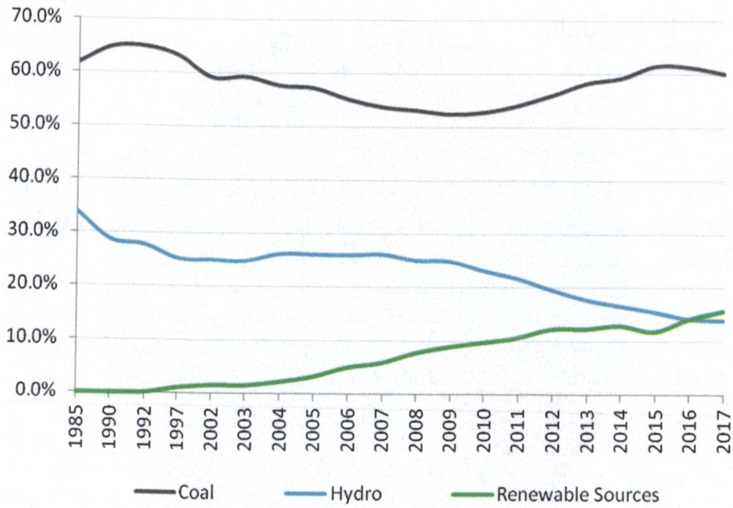

Fig. 10.1 Share of key energy sources over time in India. *Source* Ministry of Power, Government of India

When India's latest (till March 2017) energy mix is broken down (see figure below), it clearly shows the disparity between different sources. Coal dominates over others by a large margin and renewable sources come second. Given easy accessibility and other factors, the electricity produced (in absolute terms) through coal power plants will keep increasing in the medium and long run. It is projected that under optimistic scenario of high renewable energy growth, the total coal-based energy will reach 248 GW by 2022 and 218 GW by 2030 (after accounting for retiring of inefficient coal power plants). Under the low renewable energy scenario, this figure might reach 474 GW by 2030.[2] But given government's plan to scale up renewable energy in India in the next five years, it is expected that this sector will capture significant portion of the energy mix (Fig. 10.2).

In the figure presented below, the 15.6% share of renewable energy in the existing energy mix is further broken down into its components. Wind power (28.7 GW) dominates this sector followed by solar (9 GW). But the Indian government has set itself a much ambitious target of reaching 175 GW of renewable power by 2022 out of which 100 GW is planned to come from solar (utility scale, distributed, off-grid/mini-grid). This 10-folds increase in solar capacity in India in the next 5 years is a difficult goal to achieve. But as seen in the graph below, if past growth trend in solar energy is any indication, this target might not be an impossible one, through a lot needs to be done if India wants to reach close to the 100 GW mark. Between July 2016 and March 2017, India's solar capacity increased from 8 GW to 9 GW. This pace has to increase to 18 GW per year if India wants to gets to its 100

[2]ibid.

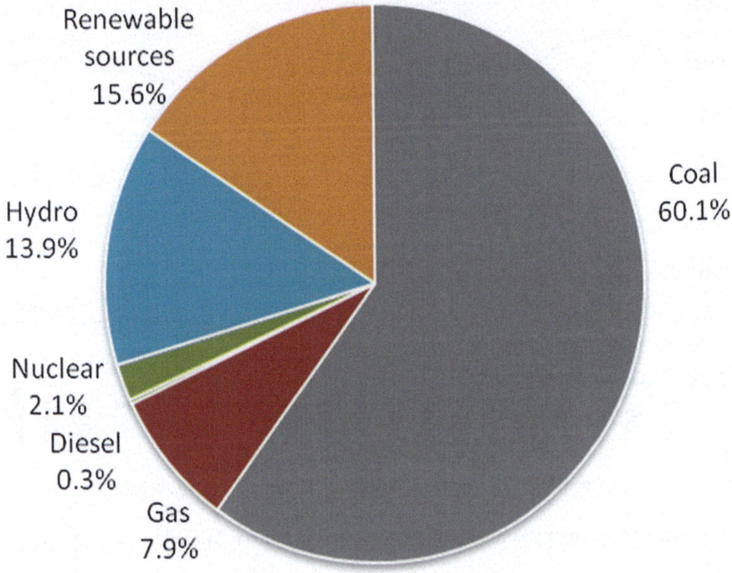

Fig. 10.2 Current share of key energy sources in India. *Source* Ministry of Power, Government of India

GW target in the next 5 years. Many experts and international agencies like International Energy Agency (IEA) has projected India's solar target to reach 40 GW by 2022 (Fig. 10.3).

10.3 India's Need for Solar and Its Solar Potential

There are many benefits of solar energy, especially for a developing country like India. The development and industrialization path is usually energy intensive with high peaks of energy demand. Meeting this demand through conventional sources like fossil fuels is costly and environmentally damaging. Besides, electricity losses in India are also very high. The latest government figures suggest that including transmission and distribution losses, the aggregate technical and commercial losses is close to 25%. The peak demand deficit has improved over the years and now stands at 0.5% for India with Eastern and Northern parts having more than 1%.[3] Electricity theft is also prevalent in most parts and by some estimates costs 1.5% of India's GDP thus hampering India's economic growth. Fortunately, solar energy is available during the daytime and it coincides with the peak demand, and adopting it

[3]Ministry of Power (2017). Government of India http://www.cea.nic.in/reports/monthly/executivesummary/2017/exe_summary-01.pdf.

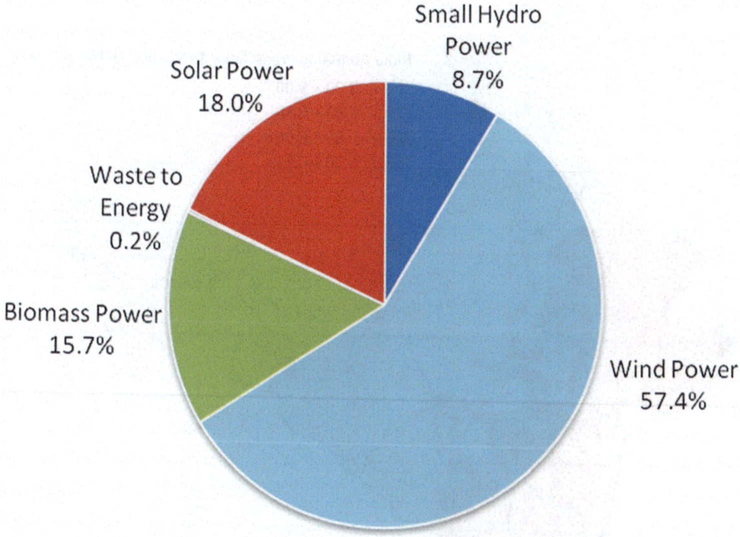

Fig. 10.3 Current distribution of renewable energy sources. *Source* Ministry of Power, Government of India

can reduce the cost to meet the peak demand for people. If solar energy plants are scaled up, there might not be any need to build additional energy generation capacity. Studies have also demonstrated that with the rapid drop in per unit solar energy cost in the last few years, it also makes economic sense for India to adopt and scale up at a faster pace. Given its nonpolluting nature, in these times of global warming, climate pacts, and India's Intended Nationally Determined Contribution, solar is an attractive option.

Solar energy is also a natural choice for India because of its geographic location. It being a tropical country receives adequate solar radiation for almost 300 days in a year. This roughly translates into 3,000 hours of sunshine, which if properly harnessed can generate 5000 trillion kWh of electricity. This solar potential in India comes from the U.S. Department of Energy's National Renewable Energy Laboratory (NREL). On the basis of a 10-year average (2002–2012),[4] the annual average irradiance data is plotted in the map below. The data indicates that most of the regions in the country receive 4–7 kWh of solar radiation per square meter thus suggesting that most part of the country is suitable to set up solar power plants. Some states like Gujarat, Rajasthan, Maharashtra, Karnataka, Andhra Pradesh, Tamil Nadu, and Madhya Pradesh are best suited for solar energy generation as most parts fall within areas with high solar potential (> 5.5 KWh/m^2/day). States like Odisha, Jharkhand, Chhattisgarh, Mizoram, Kerala, as well as parts of Bihar,

[4]National Renewable Energy Laboratory (NREL), U.S. Department of Energy http://www.nrel.gov/international/docs/readme_india_solar_maps.txt.

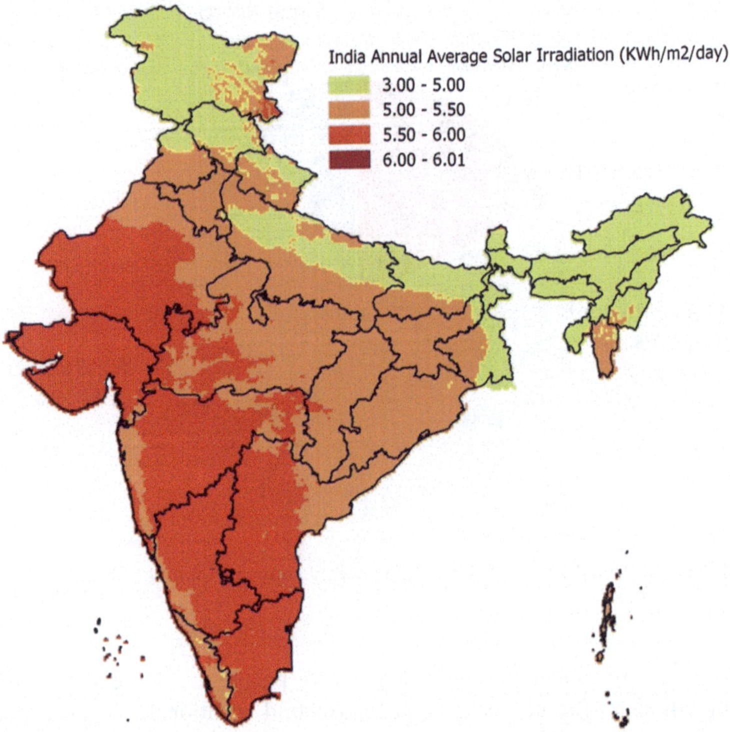

Fig. 10.4 Solar energy potential in India. *Source* Author; National Renewable Energy Laboratory, U.S. Department of Energy

Punjab, West Bengal, Uttar Pradesh, and Uttarakhand fall within the category of medium solar potential states (between 5.0 and 5.5 KWh/m^2/day), though above the threshold level needed to generate solar power energy. Many of the isolated parts in the North Eastern states, Himachal Pradesh, and Jammu and Kashmir can also harness solar power though the potential is on the lower side (Fig. 10.4).

10.4 Harvesting India's Solar Potential

Most parts of India get abundant amount of solar energy and given our national electricity demand and environmental objectives, it would be ideal to tap into the solar potential. This can be done through converting solar energy into electricity, primarily through photovoltaics (PV). But the generation of solar-based energy can happen in two conditions, i.e., grid-connected and off-grid. The former is chosen because it has low running cost, solar is available during the day which coincides with peak demand time, and people can enjoy capital return through net metering.

The latter off-grid option is also a viable choice and it is chosen because in places where power availability is low or it is too costly to bring grid, people can generate their own power through solar PV (with or without storage). It is environmentally safe and pollution free (compared to off-grid power from fossil fuels) and it also brings a factor of self-reliability. Additionally, there are no transmission losses and no need of overhead wires. Though both options have their advantages, they can be set up under different situations. The ground-mounted solar is grid-connected; rooftop solar can be both, while there are some decentralized options that are only off-grid. All these options are promoted by the government (at different levels) based on their respective targets and collectively they are set to achieve 100 GW by 2022.

10.4.1 Ground-Mounted Solar Schemes

Under the ground-mounted schemes, the primary objective is to achieve the set target through development of solar parks. Government has proposed to develop ultra mega solar park projects in the next 5 years to get to the capacity of 40 GW. In the initial break-up of solar energy targets, solar parks were assigned to contribute 20 GW but in February, 2017 this target was doubled. This increase in solar park target has not changed the overall solar target of 100 GW by 2022.

As per design, each of the solar parks developed to get to the 40 GW target will be 500 MW and above. Since these are large projects, the parks will be set up in collaboration with the respective states governments and developer for the park or the implementing agency that will be selected by the state government. One of the preconditions in this scheme is for the state to buy at least 20% of the solar power generated from these parks. So far 34 solar parks have been approved and they are in various stages of development. The process of developing these parks has faced some initial hurdles like delay in land acquisition and development of internal infrastructures including internal transmission systems by the solar power park developers. Despite such hurdles, thus far 8.9 GW solar capacity has been created through solar parks.

Besides the large solar parks, government also has other small-scale ground-mounted solar schemes. Part of them is for the defense sector, while Solar Energy Corporation of India (SECI) has been given responsibility to set up grid-connected solar power projects of 0.75, 2, and 5 GW under viability funding scheme. Government-owned power generation companies like National Thermal Power Corporation (NTPC) and National Vidyut Vyapar Nigam Limited (NVVN) are asked to set up 15 GW grid-connected solar power plants. These conventional plans are supplemented by innovation schemes like installation of solar PV on canal banks or canal tops under Nation Solar Mission.

10.4.2 Rooftop Solar Schemes

Rooftop solar is one of the key components in achieving the national solar target. The nationally set target for rooftop solar is 40 GW. A 2014 study by the National Institute of Solar Energy (NISE) has estimated a rooftop solar potential in India to be 42.8 GW. A different study in 2014 by Bridge to India has estimated this potential to be 57–76 GW from all types of buildings (small residential to commercial and industrial). Whereas another 2014 study by TERI has estimated market potential of rooftop solar in India to be around 124 GW. These studies and their findings suggest that in the long run, India can achieve much higher targets through rooftop solar. The current target set by the government will be achieved through grid-connected and off-grid-connected systems. Irrespective of the systems adopted by the users, rooftop solar is seen as an option with multiple benefits. Unlike the large-scale solar projects, rooftop solar does not require any land pooling and there is also no need for separate transmission. They incur minimal technical, transmission, and distribution losses and can help manage daytime peak load. Such systems are also beneficial for power distribution companies (DISCOMS) since by purchasing from such sources, they can meet their renewable purchase obligation for 8% of electricity consumption.

There are two development models to implement rooftop solar in India. The first is the CAP-EX model in which investment is made by the rooftop owner while the second is the RESCO model where investments are made by the developer and it gets into an agreement with the rooftop owner. At present, close to 87% of India's current rooftop solar is built on the cap-ex model.[5]

10.4.3 Off-Grid Decentralized Solar Schemes

In recent times, India's supply side of electricity has gone through significant augmentation but some parts of the country still live in dark and await electrification. To address this gap, various decentralized PV systems are supported under this program. Maximum capacity per site is 500 kW and it mainly supplement lighting and electricity. Mini-grids for rural electrification are also covered under this scheme. On the smaller scale, solar lighting and solar pumping are also covered. As these plans are targeted in the poorer areas, much financial help is provided by the government. Capital subsidy of 30% is provided to the end user and the subsidy goes as high as 90% for special states.

[5]Bridge to India, 2016. http://www.bridgetoindia.com/rooftop-solar-market-in-india-witnessing-rapid-growth-but-2022-target-seems-elusive/.

10.5 Steps Towards Universal Energy Access and Energy Security

The NAPCC is a landmark document for India's energy future as two out of eight national missions outlined are directly linked to energy. The first was the National Solar Mission (renamed as Jawaharlal Nehru National Solar Mission (JNNSM)) and the second one was National Mission for Enhanced Energy Efficiency. Here, the primary focus will be on the former mission as it was framed to promote power generation from solar energy while also integrating other renewable sources like biomass and wind with solar energy. Under the solar mission, both solar thermal and solar photovoltaic are suggested to capture solar energy. As it is a relatively new technology, the government also plans to focus on capacity building so that condition is created for scaling up and increased solar energy penetration in the country. Solar manufacturing is also encouraged within India. On the consumer and business side, for a faster uptake of solar energy, the solar mission also outlines various tax exemptions, capital subsidies, and incentives for various components in solar energy value chain. The center provides 40% accelerated depreciation income tax for commercial and industrial owners, 30% capital subsidy for residential and institutional owners, while some states supplement these with generation-based incentives (GBI). These incentives are discussed later in this chapter in detail. Manufacturers of certain solar products like solar lanterns, street lights, blinkers, and traffic signals can also avail capital subsidy. Along with these, assembly of solar modules using imported cells is promoted. The import of solar cells is free from import tax.

Besides the Solar Mission, the government also wants to tackle the issue of universal access to electricity through other policies and programs. Many of these programs capitalize on the India's solar potential to achieve the goal of universal access to energy. One such policy is National Rural Electrification Policy of 2006. As more than 40 million rural households are under-electrified, this policy planned to provide access to quality and reliable electricity at minimum lifeline consumption of 1 unit per household per day per year by 2012. These households usually reside in remote villages with off-grid solar-based electricity supply is more feasible. Rajiv Gandhi Gramin Vidyutikaran Yojana is another program taken up by Rural Electrification Corporation but it focuses on bulk power purchase and management of local power distribution through standalone systems. The villages that are not covered by the above program fall under the Remote Village Electrification Program. The particular technology for village electrification is chosen by state agency, based on technical feasibility and resource availability. Since, solar energy is abundant in most parts of India; it is good and reliable technology to power these remote villages. Under the Renewable Energy Supply for Rural Areas scheme, the objective is to develop and demonstrate commercially viable models of decentralized energy supply in rural areas from renewable energy.[6] Since urban areas are sites of high energy demand, the government has introduced programs for some of the large settlements. Ministry of New and Renewable Energy runs the Solar/Green Cities program under which 60 cities are developed as solar cities with the objective of meeting at least 10% of the total energy demand from renewable sources, with primary focus on solar technology. But the government understands that achieving universal access and energy security while meeting the solar target would not be an easy task. It will require much more than few national level programs. Hence, the central government has introduced many financial incentives for varied types of stakeholders and earmarked budget to promote solar in India.

10.6 Fiscal Assistance and Incentives

The solar target in specific and the renewable energy targets in general within the time frame as set by the government are ambitious to say the least. The uptake of solar energy though has picked pace in the last few years, it has not gather the momentum to get to the scale required to meet the set target. One of the hurdles to this end is the upfront cost required for installation (at the household level) and availability of sizable capital for large customers (for utility scale and for industrial and commercial uses). Though many of these limitations still exist, the government has introduced multiple fiscal incentives so as to promote solar energy and achieve much of the target. Some of these measures are discussed in this section.

The total solar energy target by 2022 is 100 GW and out of which 40 GW is supposed to come from solar parks and 40 GW from rooftop solar. In terms of financial assistance for solar parks, the total allocation set aside for them is around INR 81 billion.[7] This fund is distributed under the Central Financial Assistance program and will provide INR 25 lakh per park to prepare the detailed project report. For subsidy on power generation, INR 20 lakh per MW or 30% of the project cost (whichever is lower) will also be provided. As per the government figures, considerable portion of the target has been set up or sanctioned. On the other hand, achieving the rooftop target is more challenging given its much decentralized nature and other technical issues. Under the National Solar Mission, the government has set aside INR 5000 crore for implementation of grid-connected rooftop solar systems till 2020 across the country. According to government reports,[8] this allocation will support 4.2 GW of rooftop solar. To encourage households and institutions to adopt solar rooftop, central government provides a

---

[6]Central and State Government Solar Policy, Energy Alternatives India (EAI). Accessed from http://www.eai.in/ref/ae/sol/policies.html.

[7]Bridge to India (2017). India Solar Book, 2017.

[8]New and renewable energy sector: Achievement Report. (2017). Department of Industrial Policy and Promotion, Ministry of New and Renewable Energy. Government of India.

30% capital subsidy for most of the states while this subsidy goes up to 70% for Special category states (North East, Uttarakhand, Himachal Pradesh, Jammu and Kashmir, Lakshadweep and Andaman and Nicobar islands). These subsidies will be given out till the cumulative subsidy amount reaches INR 12,000 crore.[9] To encourage solar rooftop at the local level, many state governments are supplementing the central government subsidy with their own schemes. For example, in the state of Gujarat, an additional INR 10,000 per kW is provided to households, while in Delhi, subsidy of INR 2 per kWh is given out through Generation-Based Incentive (GBI) model. Such measures will help in achieving state's respective targets set by the national government. For industrial and commercial users, government's approach is different. Given the cost of installation and thus the cost per unit of solar power has decreased significantly in the recent past, government does not see the viability of extending capital subsidies to other uses. As a result, the capital subsidies for households are not extended towards industrial and commercial establishments but they are eligible to claim appreciate depreciation. For such properties, around 35% tax is calculated on the accelerated depreciated rate of 40% on of the solar system asset. Another approach taken by the government to ease financial burden of setting up solar rooftop is through bank loans. Renewable energy was added to the list of priority sector lending category by Reserve Bank of India in 2015. Under this guideline, all banks are encouraged to give loans for solar rooftop and now its cost can be added as part of the housing loan. The loan size is limited to INR 10 lakh for individual households.[10]

As discussed earlier, while many of the programs have a national scope, the Solar City Mission is one of the dedicated programs for urban settlements. Under the Central Financial Assistance (CFA), up to INR 2.50 crores will be provided to each pilot solar city for any renewable energy project or device installation. This is a conditional fund and is only provided if the same amount can be made available by the city administration or by the state. Additionally, up to INR 9.50 crores will be provided through CFA to the solar city again with the condition that same amount has to be arranged by the city administration/district/state on their own or other sources including PPP. Besides these hard grants, the central government is willing to commit up to INR 4 lakh for seminars/workshops trainings, awareness campaigns and a city can conduct up to 50 such events and maximum amount for these activities is restricted to INR 1 crore. It is clear from the prescribed distribution of funds that central assistance for meeting solar target is equally important to making local government financially self-reliant and enterprising along with building capacity for solar technology.

Beyond the earmarked funds, to continue a steady stream of capital for renewable energy projects, recently the Clean Environment cess on coal, lignite, and peat has been doubled from INR 200 per ton to INR 400 per ton. This increase

---

[9]Report of the Expert Group on 175 GW Renewable by 2022. 2015. NITI Aayog, Government of India.

[10]Priority sector lending-targets and classification. 2015. Reserve Bank of India.

in cess rate was introduced in the 2016–17 Budget. The impact of this has been significant as the total amount collected from across the country has increased more than 4-time in the last few years. The total collection was INR 3217 crore in 2013–14, which increased to INR 13,848 crore in 2015–16. In 2016–17, the total collection is estimated to reach INR 23,944 crore.[11] This cess has not only helped the government to raise capital for environmental issues including renewable energy generation, by increasing the cost of coal, it has made solar power more competitive.

10.7 Growth of Solar and Its Future Challenges in India

The graph (Fig. 10.5) below shows the accelerated growth of solar as a share in India's total renewable energy installed capacity. The figures for 2017 indicate that share of solar has reached above 22% of total installed renewable energy.[12] The growth of solar energy in the last decade started with government's ambitious solar target of 100 GW and thus far more than 14 GW has been installed with close to 90% coming from utility scale solar projects. This uptake of solar has been a very important story in India's energy generation journey. Some of major factors for its fast growth and high uptake are addition of mainly utility scale projects, issuing of new tenders for such projects, reduction in equipment price, and financial strengthening of electricity distribution companies (DISCOMs) through the UDAY scheme. Significant capital subsidy and tax incentive for solar energy systems at the disaggregate level have also played an important role in this journey.

The proliferation of utility scale solar projects can be attributed to some big players coming into the market through merger and acquisition. These players include TATA Powers, Geenko, IDFC Alternatives, and Azure.[13] Intense competition in the market has led to sharp drop in solar tariff in recent months with it reaching INR 3.15/kWh for NTPC's 250 MW project in Kadappa, Andhra Pradesh[14] (Fig. 10.6). With such low rates, solar energy has achieved parity level to grid and other sources. Such rates have led to high solar power demand from the DISCOMs (though the total power demand remains weak through the country).

[11]Economic Times, August 11, 2016. http://economictimes.indiatimes.com/industry/banking/finance/government-may-get-rs-23944-crore-from-clean-environment-cess-in-fy17/articleshow/53651980.cms.

[12]Installed Capacity-June 2017, Central Electricity Authority, Ministry of Power (2017). Government of India.
http://www.cea.nic.in/reports/monthly/installedcapacity/2017/installed_capacity-06.pdf

[13]2016 was a great year for the Indian solar industry but the best is yet to come (2016). Bridge to India.
http://www.bridgetoindia.com/2016-great-year-indian-solar-industry-best-yet-come/.

[14]More to Indian solar tariffs than meets the eye (2017). Bridge to India.
http://www.bridgetoindia.com/indian-solar-tariffs-meets-eye/.

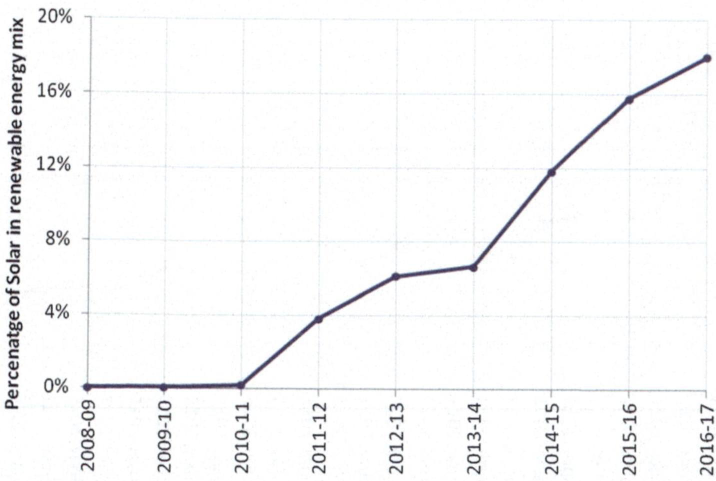

Fig. 10.5 Share of solar energy in renewable energy mix over time in India. *Source* Ministry of Power, Government of India

The government has larger plans for solar parks like the plan to build 25 Solar Parks and Ultra Mega Solar Power Projects in the next 5 years with a total target capacity of 20 GW.[15] As for decentralized rooftop solar power is concerned, government provides attractive capital subsidies which have resulted in greater than 100% year-on-year growth between 2015 and 2016. As of 2017, the total grid-connected rooftop solar capacity has exceeded 1 GW. The demand and uptake for small off-grid home solar systems are also envisioned to grow given the remote areas of India still suffer from poor or no electricity whereas urban areas suffers from frequent power cuts.

Through the recent trends for the growth of solar energy in India are encouraging, the sector faces some challenges moving forward. The above discussion on falling prices of solar power tariff on one hand shows the completion in the Indian market, keenness of developers to get in and invest, and incentives provided by the government while it also indicates towards uneven playing field that is being created by cheap imports of solar cells and modules. The foreign cell manufacturers (mainly in China and Malaysia) often enjoy state subsidies, which leads them to practice dumping and predatory pricing. Chinese solar product makers can afford low prices in India also because they face overcapacity in their country and export to Europe comes with steep duties.[16] These factors have disrupted the domestics market and effectively hurting the Indian solar manufacturing sector the most.

[15] Fact Sheet on Scheme for Development of Solar Parks and Ultra Mega Solar Power Projects (2016), Ministry of New and Renewable Energy, Government of India.
http://pib.nic.in/newsite/Printrelease.aspx?relid=145542.

[16] Down to Earth, Center for Science and Environment.
http://www.downtoearth.org.in/blog/what-ails-india-s-solar-energy-drive–58134.

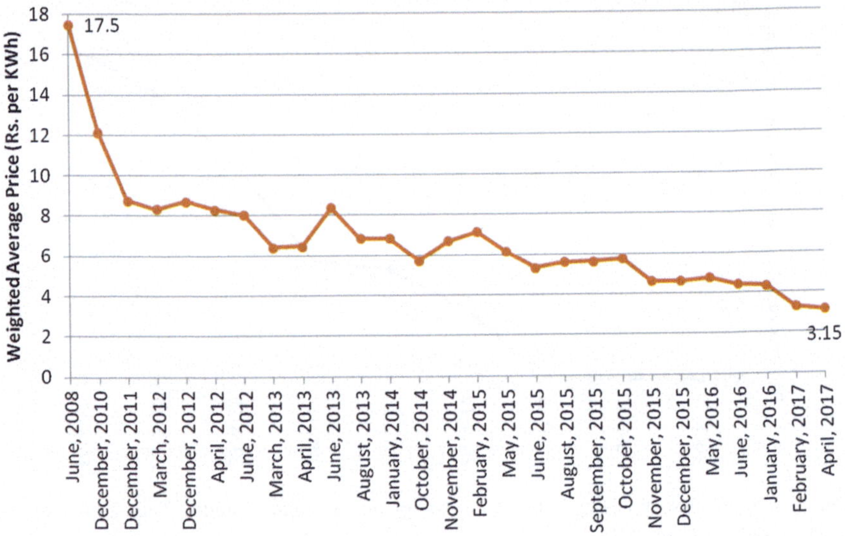

Fig. 10.6 Average per unit price of solar energy over time in India. *Source* Ministry of New and Renewable Energy, Government of India

The scale of disruption can be imagined by the fact that 85% of solar sector depends on the cheap imports.[17] As a result, some of the biggest solar equipment manufacturers in India are facing financial crisis due to such competition. In the near future, as the government moves towards its solar target and as the monetary stakes go up, such large dependency on cheap imports can cost India heavily. Thus, it is important that the government creates a level playing field for Indian producers by revising its solar policy. This will also help in checking the free fall of power tariff, which is needed for a healthy growth of solar energy system in India. So far, to arrest the fall in tariff, government has slowed down its procurement policies leading to holding back or postponement of new utility scale tender announcements and project allocations.

India's overall solar target is very ambitious by any standard. But given the trend of the utility scale projects, its share might well be achieved. Where India lags is the rooftop solar targets. Out of 40 GW only 1.2 GW has been met as per the latest figures in 2017. Because of this sluggish pace for rooftop solar uptake the overall national solar target seems difficult to achieve. There are various reasons that have been cited for the poor uptake including long bureaucratic procedures and unfamiliarity with the process of getting rooftop solar installed. Though government at various levels is promoting it, people still do not have full knowledge of government's capital subsidies and incentives, they fear of large upfront capital

[17]ibid.

requirement and are uncertain of return-on-investment ratio. Even if an educated consumer buys into the rooftop solar technology and its related cost, there is a dearth of trained personnel to deliver the complete product and process. At the policy level, ineffective implementation of net metering in various states has not helped the cause either. As for the developers, lack of uniform and continuous rooftop space in most settlements makes their task difficult to install large solar panels.

Some of the other problems that have marred this sector are its inherent nature of large land requirements for ground-based projects. To install panels to generate 1 MW of power, 2.5 acres of land is required and this figures goes up to 4 acres when all the space required for other accessories is considered. Such large land parcel adds to the cost of such solar projects. On the technical side of the matter, injection of large solar generated power can disturb the grid stability. As the amount of electricity from solar farms is not continuous and it can be difficult for grid operators to predict the precise input at any given hour to maintain grid frequency, a small error can lead to instability of the grid. At the environment front, studies have shown that manufacturing of solar cells is an energy-intensive process, which is often derived from burning fossil fuels. So, though there is no carbon emission during electricity generation from solar energy, emissions are generated during various stages of the photovoltaic lifecycle. All these challenges need to be addressed in some form and to some extent so that the shadow of doubt over solar energy in India can be removed.

10.8 Conclusions

In the last two decades, the rise of renewable energy within India's energy mix has been considerable and meaningful. For the first time, renewable energy sources have surpassed hydro-based power in India but it still lags coal power. This growth, especially in recent years, has been the result of government's push for universal energy access, energy security, and addressing its environmental and climate change concerns. These efforts have manifested in the form of 175 GW target of renewable energy by 2022. Solar energy has received much attention among the renewable sources and hence the target for this sector alone is 100 GW. Such ambitious target has rightly captured the solar energy potential in India and harvesting this potential can lead to the promised goal of universal access and energy security in the future. To this end with solar target as the means, many plans, programs, and policies have been rolled out in the last few years. Significant amount of capital resources have been earmarked for solar projects while financial incentives in various forms are given to the users. These efforts have yielded dividend with high uptake of solar energy with continuously falling solar tariff rates. Despite recent success, some challenges still remain at the central and state policy as well as local implementation levels. These hurdles need to be addressed soon so that the growth of solar market in India can be a healthy one.

If the government can balance its act to push solar energy in India while maintaining a healthy competition in the market, it can reach its set goal and when it does, its energy future will indeed look sunny.

Acknowledgments This work was conducted in his time at the Indian Council for Research on International Economic Relations (ICRIER) and does not represent Rio Tinto's views.

Chapter 11
Water, Ecosystem Services, and Food Security: Avoiding the Costs of Ignoring the Linkage

Nilanjan Ghosh

Abstract This paper talks of the emerging paradigm of water management that acknowledges critical ecosystem services, and challenges the linear and positive relation between water availability and food security. The ways water used to be managed, globally, are changing rapidly. The existing engineering modes of water management entail constructing large structures intervening into the natural hydrological flows, and exploiting the water for human use. A large component of demand for water emerged from the need of the agricultural sector in various parts of the developing and developed world to ensure food security. Over time, the developed nations began realizing that such traditional engineering ways of water management entailing large constructions are not sustainable in the long run, and can have serious impacts on ecosystems. Since large parts of livelihoods are dependent on the ecosystem services, negative impacts on ecosystems affect livelihoods negatively, too. Hence, a new paradigm of water management recognizing the ecosystems livelihoods linkages is emerging. This new paradigm is known as Integrated Water Resource Management (IWRM) and, when applied at the level of a river basin, is referred to as Integrated River Basin Management (IRBM). This new paradigm delinks economic growth and food security from increasing water use, and provides for an ecosystemic definition of food security. However, this changing paradigm is yet to be recognized in policy documents of the developing world, especially India. For India to embark upon a low-carbon growth trajectory, it must embrace the new paradigm of water management.

N. Ghosh (✉)
Observer Research Foundation, New Town, Kolkata, India
e-mail: nilanjan.ghosh@gmail.com

N. Ghosh
WWF-India, New Delhi, India

R. Kathuria et al. (eds.), *Low Carbon Pathways for Growth in India*, India Studies in Business and Economics, https://doi.org/10.1007/978-981-13-0905-2_11

11.1 Introduction

The fact that ecosystem services and food security are inextricably linked is being increasingly recognized within academic circles, even though it rarely finds reference in the developing world's policy documents. In South Asia, this omission has led to adherence to the archaic notions of water management entirely based on the reductionist engineering paradigm looking at short-term economic benefits, and that ignores long-term social and ecosystem concerns. This paradigm is essentially an integral component of the colonial legacy as this was introduced and formalized under colonial capitalism in South Asia leading to a "metabolic rift" between human–nature relationship (Foster 2003; Gilmartin 1994, 1995). The most critical concern that the reductionist engineering paradigm misses addressing is that the livelihoods of the poor in the developing world are reliant on ecosystem services. Essentially, because of the importance that ecosystem services render to the livelihoods of the poor, such services are often classified as "GDP of the poor" (Martinez-Alier 2012). Unfortunately, India's policy documents and implementation plans rely on "arithmetic hydrology" rather than "eco-hydrology" and have ignored this linkage. They have also ignored the changing relation between water and food security, with the change being embedded in the new emerging paradigm of water management, also known as Integrated Water Resource Management (IWRM), which recognizes the critical role of ecosystems.

This paper therefore attempts to present the changing relation between water and food from the perspective of IWRM, where the ecosystem is considered an important component of water demand. The paper also highlights how water policy documents in India have ignored the notion of Integrated River Basin Management (IRBM).

This paper consists of seven sections. Section 11.2 of this paper relates the linkage between ecosystems and food security. It highlights the fact that food production is a provisioning service of the ecosystem, and therefore the ecosystem plays an important role in long-run food security. Section 11.3 talks of conflicts over water and land use arising from economic (agricultural) and ecosystemic use. It also talks of how dam construction (with irrigation as the major purpose) leads to conflicts over water use in India; it also brings in the debate over river interlinking. Sections 11.4 and 11.5 talk of the tenets of Integrated Water Resource Management (IWRM) and Integrated River Basin Management (IRBM), respectively. Section 11.5 also talks of how water policy documents in India have missed taking a river basin approach in the context of water resource management. Section 11.6 talks of the changing relation between water and food, and attempts to present an ecosystemic definition of food security. It is here that I explain how adherence to IWRM and IRBM provide pathways towards low-carbon growth. Section 11.7 consists of the concluding remarks.

11.2 Ecosystems and Food Security

Of the entire range of services provided by ecosystems (provisioning, regulating, supporting, and cultural) to human society (MA 2005), food provisioning, either naturally or through human intervention, is one of the most important. As pointed out by Richardson (2010), the role of ecosystem services in enabling food security needs to be looked at from three aspects, availability, access, and utilization of food. The structure of the ecosystems supports these utilities, through provision of critical ecosystem services facilitating production of food, creating opportunities to generate incomes, and creating a natural base for provision of energy for cooking (Richardson 2010).

As such, agricultural systems fundamentally depend on ecological processes, which clearly explain the production aspect. What is less understood is the role of ecosystem services in ensuring access to food. Sen (1981) postulated that food security cannot only be a function of availability, but also must be a function of access. Household-level access to food is facilitated and supported by ecosystem functions, directly or indirectly. These include provisioning services that allow for the transport and processing of food as well as for the production of agricultural goods and raw materials that can be sold to generate income. One of the most critical examples in this regard is the creation of nonfarm employment opportunities that help generate incomes for households (Richardson 2010). Households in the rural areas of the developing world engage in harvesting and use of wood and non-timber forest products (NTFPs) which often emerge as another source of their livelihoods, enhance their purchasing power, and increase their access to food, and nearly one-third of the world's forests are primarily used as a source for such products. Given the seasonal nature of agriculture, the production and sale of charcoal, food, and other NTFPs is important in sustaining many rural households during the off-season (Osemeobo and Njovu 2004; Richardson 2010).

The utilization dimension of food security is concerned with how households utilize the food accessible to them. Therefore, while access is a necessary condition of food security, it cannot really be the sufficient condition till the utilization criterion is satisfied. Utilization is generally a function of safe and sanitary cooking practices and the quality of nutrition (Webb et al. 2006). Ecosystem services contribute to the utilization of food by households and smallholders in various ways. These might occur through the supply and availability of safe drinking water and food preparation; the fuels and energy for hygienic heating, cooking, and storage of food; the materials for sanitation and health care; and the micronutrients necessary for an adequate diet (Richardson 2010). Safe and healthy cooking of food is a crucial component of food utilization: this helps in improving the nutritional value of food, preventing disease, and enhancing the taste. Biomass sources are used in various parts of the developing and underdeveloped world for energy needs of cooking. A large part of this is fuelwood, lops and tops, and NTFPs (Richardson 2010). Nature further provides add-on spices that enhance taste, and add to the nutritional quality of food (Richardson 2010).

11.3 The Conflictual Outcomes

Agricultural expansion during the last century has caused widespread changes in land cover, watercourses, and aquifers, thereby degrading ecosystems, and restricting their ability to support some services including food provisioning (Falkenmark et al. 2007). Agricultural expansion in most of what are perceived to be the "water-scarce" economies were essentially results of intervention in hydrological flows through constructions of large dams and storage and diversion mechanisms (Ghosh 2009). No doubt, making more water available for irrigation allowed water-intensive crops to be grown and enabled land-use change but the latter has threatened the ecological foundation of the world food system. Quite unfortunately, the management policy of many agro-ecosystems has essentially been based on the premise that they are delinked from the broader landscape (Falkenmark et al. 2007). There has been scant recognition of the ecological components and the processes that support the sustainability of such agro-ecosystems. As a result, the carrying capacity of the ecosystem has been defied by traditional agricultural and water management regimes. Some ecosystems, therefore, were made to cross the ecological thresholds, leading to a regime change in the ecosystem and their concomitant services (Falkenmark et al. 2007). The resultant reduction in the ecosystem's resilience also restricts the sustainability of its food provisioning service. Unfortunately, beyond a point, even the water supply augmentation plans (through dam constructions), and land-use change (entailing bringing more land under agriculture by cutting down forests or filling wetlands), do not work and can have a negative impact on food security, with the impacts intensified by climate change, as argued by Chaturvedi (2015) in this volume.

Threats to the ecological foundations of agriculture arise from resources that are becoming scarce over time, because of increasing competing uses that are getting diversified in nature, and increases in the human demand for food and other uses due to population growth and changes in human preferences, thereby validating the Malthusian creed. The drivers of this process are: competition for land and water, traditional resource-consuming agricultural practices, deforestation, and unsustainable pesticide use (that reduces the long-term soil productivity, and also contaminates groundwater), and climate change. This accentuates conflict over water and land. Poor people in the developing world, who rely on ecosystem services for their livelihoods, are extremely vulnerable to ecosystem changes. Therefore, there is no doubt that the failure to tackle ecosystem degradation and loss can severely undermine the attempts towards achieving the Millennium Development Goals (to be replaced by Sustainable Development Goals or SDGs after 2015) of poverty reduction, food security, and environmental sustainability.

Water conflicts, more often than not, have been results of a constructionist paradigm. The western world, led by USA, was the harbinger of development through this constructionist regime. From the 1920s to 1960s, huge dams were constructed, more so for irrigation needs. But, over time, serious ecological impacts resulted over the Colorado River basin in the western US, for example, the construction of the Hoover Dam, the Tennessee Valley Project, and the Central Arizona Project have led to ecological problems, whose long-term costs are higher than the short-term benefits. Environmentalists have been vocal about the livelihoods problems that have been an outcome of the ecosystem damage through the losses in ecosystem services. This has led to a trend in the western world to decommission dams. As noted by Gleick (2000), around 500 dams have been decommissioned in the US and Europe in the 1990s, noting the extensive ecosystem damages and potential for conflicts. Rehabilitation and livelihoods losses even during the construction phases have been sources of social conflict (Homer-Dixon 1994), the costs of which are often not taken into consideration while carrying out impact assessment (Ghosh 2008).

Hence, the outcome is not merely a conflict between communities and government agencies but it is a conflict over sectoral use of water as well, with the ecosystem emerging as a critical source of water demand from the sustainability perspective. This is increasingly being recognized by the emerging paradigm of Integrated Water Resources Management (IWRM).

11.3.1 Supply-Side Interventions and Water Conflicts in India

Ghosh and Bandyopadhyay (2009) postulate that one of the major reasons for water conflicts in India have been water supply augmentations plans. This was shown by them in the context of the Cauvery River basin, where they find evidence of how attempts to reduce the "scarcity value" of water by plans to augment supply for paddy cultivation intensify conflicts between the riparian states of Karnataka and Tamil Nadu. It is further argued that one of most important reasons for the conflict is that irrigation water is highly subsidized and is therefore treated as a "zero-value" resource. A free resource is prone to be wasted, and that is exactly what prevailed in the Cauvery basin, leading to conflicts. Ghosh (2015) further goes on to argue that the environmental security concerns over the transboundary water relations between India and Bangladesh have arisen more due to the reliance on the reductionist engineering paradigm brought into South Asia by British engineers, who hardly had much idea about waters flowing down the Himalayan terrain. The application of "one for all" technology in water resource planning and management has been the prime cause of concern.

11.3.1.1 Proposal for Interlinking of Rivers in India

The proposed River Link Project (RLP) in India is based on traditional engineering perspectives, and one of the latest glaring examples of the reductionist "arithmetic hydrological" paradigm based approach to water management in South Asia

(Ghosh 2012). It is a very large project for storage and long-distance transfer of water, mainly from the Ganges-Brahmaputra-Meghna (GBM) basin to river basins in drier areas in western and southern India (Fig. 10.1). The project includes the construction of nine large and 24 small dams and digging of 12,500 km. of canals. This project has drawn serious criticism from the perspective of sustainability and equity (Bandyopadhyay 2009: 147–183) and from that of economics (Alagh et al. 2006). Bandyopadhyay and Perveen (2008) have expressed their apprehensions on the interlinking of rivers project and feel that the project may further aggravate interstate water disputes, as well as aggravate the international hydro-political situation in South Asia (Ghosh 2012). They identify avenues through which new interstate conflicts may emerge. It is a fact that the federal states in India have always enjoyed rights over water for apportionment and allocation. However, under the centralized scheme of allocation under the ILR, the existing modes of riparian rights of the states get disturbed, leading to conflicts; already a few states have expressed their dissent. Unfortunately, these views, critical of the scientific credibility of such a large project, have not had any impact on the official policy. Hence, the question remains whether the official approach will continue to follow the reductionist engineering perspective or be willing to accept the emerging holistic perspective of ecological engineering (Ghosh 2012).

11.4 The Paradigm Debate Over Water and Emergence of Integrated Water Resource Management[1]

Traditionally, water has been looked at as a resource occurring in "abundance" in nature, and hence, increasing demand was never seen as posing any potent threat. Hence, the impression that became predominant, emanated from the idea that water scarcity is spatial, and more water can be diverted to the water-scarce zones from water-rich zones, through appropriate supply augmentation plans. In order for "water to be distributed equitably", the colonial engineering thought process led the idea of supply expansion plans through interventions in the natural hydrological flows (e.g., Rao 1975). As a result, water resource planning generally relied on linear projections of future populations, per capita demand, agricultural production and levels of economic productivity (Gleick 2000).

Towards the middle of the last century, serious concerns were expressed on the long-term wisdom of following such a strategy that is focused exclusively on increasing interventions into the hydrological cycle. Despite its impressive short-term successes in providing larger supplies, it is increasingly being realized that addressing the new and emerging challenges is no more possible in the long term, unless some fundamental changes take place in the way humans have looked

[1]This section draws largely from Ghosh (2008), (2012), and (2015), where I have previously talked of the paradigm debate earlier.

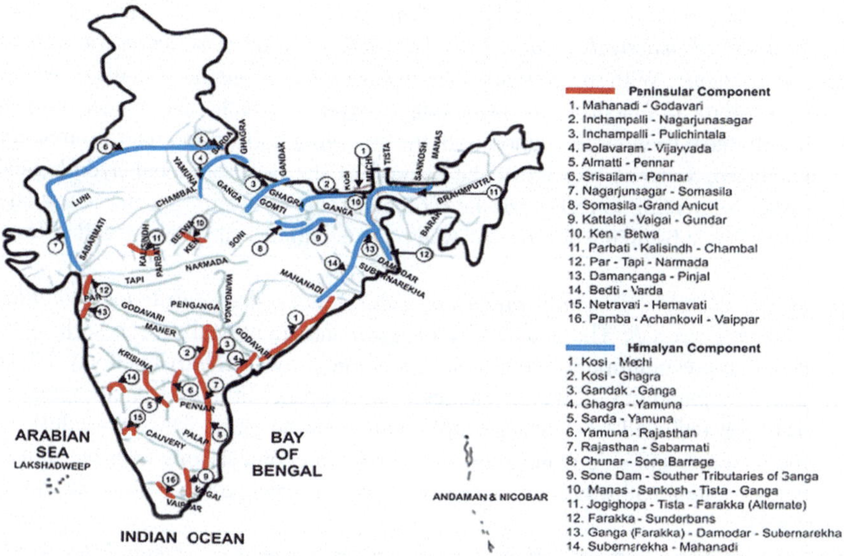

Fig. 1 Map of ILR: Peninsular and Himalayan components. *Source* Amarasinghe (2012)

at water resources so far. The "business-as-usual" thinking has started to be feared as counterproductive. There emerged the need for a fundamental change in terms of a new interdisciplinary paradigm that has been constantly gaining ground over the years. The new ways of managing water on the basis of a holistic knowledge base have increasingly been identified as Integrated Water Resource Management (IWRM).

11.4.1 The Emerging Paradigm of Integrated Water Resources Management

The professional and scientific views of water resource management are changing rapidly, based on scientific analyses of past mistakes and availability of new information. This "changing water paradigm" (Gleick 1998; Bandyopadhyay 2004) represents a real shift in the way humans think about water. The realization of a need for holistic modes of water management has been reflected in some of the policy actions of the developed world, primarily with the dawning of the ecological concerns (Gleick 2000). The new paradigm recognizes human society as a subsystem of the biosphere in which water is a key element (Falkenmark 1997; Falkenmark 2003). Based on the various contending thoughts and ideas, the notion of IWRM has been conceptualized in the form of the following points:

(a) *Water is viewed as an integral part of the global hydrological cycle, and not as a stock of material resource to be used for the satisfaction of human requirements*: With the continued emphasis on the economic benefits of water, its ecological functions in sustaining ecosystem health, and thence human health, have been largely ignored. In the emerging holistic and interdisciplinary paradigm, water is viewed in the context of the broader global hydrological cycle. Neglecting to recognize the ecological cost of diverting water is actually internally subsidizing the use of water for economic purposes at will (Flessa 2004).

(b) *Supply of ever-increasing volumes of water is not a prerequisite for continued economic growth.* The availability of water has traditionally been seen as an essential precondition for continuing economic growth (Bandyopadhyay 2004). The new paradigm, however, suggests the opposite, in that, economic growth has been delinked from water supply augmentation plans. This helps shift the focus to demand-side management of water, an approach long overdue. It also helps create a pathway for low-carbon growth (Gleick 2000; Falkenmark et al. 2004).

(c) *Clear and strict prioritization of various types of needs and demands for water, including those by ecosystems, is needed.* The new and interdisciplinary paradigm prioritizes the various competing uses of water; one is between the needs of the ecosystem and the needs of human society. The other is among the needs of human societies themselves (Bandyopadhyay 2004). An important component of current water resource management is setting the right priorities by understanding the involved trade-offs.

(d) *There is a need for comprehensive assessment of water development projects within the framework of the full hydrological cycle.* A crucial element of the new and holistic paradigm is the creation of an interdisciplinary knowledge base able to offer nonpartisan and comprehensive assessments of the justifications and impacts of water resource development projects (Bandyopadhyay 2004; Barbier and Thompson 1998).

(e) *A transparent and interdisciplinary knowledge base for understanding the social, ecological and economic roles played by water resources is required.* The complexities of managing water-related problems include a real understanding of the nature of water resources and their complex links and interrelations with other systems. This means that single-disciplinary approaches will no longer work and new, innovative strategies will have to be developed for coping with water problems, involving multidisciplinary approaches (Falkenmark et al. 2004; Bandyopadhyay 2004).

(f) *Droughts and floods are to be visualized in the wider context of the ecological processes associated with them.*

(g) *Appropriate new social and economic instruments for promoting careful and efficient uses of water resources or for the reduction of damage to their quality from pollution should be developed.* The new paradigm emphasizes the need for a new economic perspective evaluation of water. The question of pricing of water, the desirability, or otherwise, of the growing trend towards privatization of water resources as the final solution, the ecological economic valuation of the ecosystem services provided by water systems, are all part of a rapidly emerging knowledge base of water economics.

(h) *There is a need to accept restructuring the institutional frameworks for water resource development at local, state, river basin and national levels for making it equitable, sustainable, and participatory.*

These elements should be seen as indicative and not exhaustive. They are subject to further refinement as the process of the shaping of a new paradigm progresses. Such a list, for the time being, can offer the fundamental guidelines for putting the new paradigm into force. Given the above, the new emerging paradigm recognizes that irrigation development has often come with a high environmental price tag (Molden and Fraiture 2004). The costs range from degradation of aquatic ecosystems, fragmentation, and desiccation of rivers, and drying up of wetlands. Barbier and Thompson (1998) and Acreman (2000) show that in many cases the monetary values generated by irrigation proved to be less than the monetary values generated by the ecosystems they replaced. Falkenmark (2003) stresses that by benefitting from the shared dependence of humans and ecosystems on water, IWRM can integrate land, water, and ecosystems and promote the three E's—two human-dependent ones (social equity and economic efficiency), and one related to the ecosystem (environmental sustainability). As an unbiased catalyst for reconciling these concerns, and prioritizing the competing ends, valuation of the economic vis-à-vis environmental uses of water becomes critical.

11.5 The River Basin as the Planning Unit: Evolution of Integrated River Basin Management

While Integrated Water Resources Management (IWRM) became the key mantra, it was thought that the river basin should be considered the spatial unit of riverine management. This led to the development of the notion of Integrated River Basin Management (IRBM), leading to a paradigm shift from the earlier reductionist notion of project-based approach to river basin management. The primary tenet of IRBM is that naturally functioning river basin ecosystems, including any wetlands and groundwater, are an integral part of the water system. Hence, while the entire river basin is treated as an ecosystem, management of the river basin has to include maintenance of ecosystem functions and services so as not to cause destructive impacts on the ecosystem services (Boelee 2011; Mattas et al. 2014). This "ecosystem approach" is the key ideas far as the Convention on Biological Diversity (CBD 1992) is concerned.

Interestingly, over time, many policy documents began acknowledging ecosystem concerns without really understanding how to interpret them. The National Water Policy of India also acknowledges this notion but shows little application of it. For example, the 2007 Award by the Cauvery Water Tribunal, certain quantities of water are stated as being "unavoidable escapages to the sea" (sic.). In many cases, there is a clear misinterpretation of the notion of environmental flows without much understanding of the eco-hydrological processes associated with it. Environmental flows describe the quantity, timing, and quality of water flows required to sustain freshwater and estuarine ecosystems and the human livelihoods and well-being that depend on these ecosystems; through the implementation of environmental flows, water managers strive to achieve a flow regime, or pattern, that provides for human uses and maintains the essential processes required to support river ecosystems at an agreed sub-pristine level. However, most policy documents in South Asia place an ad hoc quantity or percentage as "flows" that have very little ecosystemic and scientific basis. This becomes clear in the National Water Policy 2012, Govt. of India, which states, "…A portion of river flows should be kept aside to meet ecological needs ensuring that the low and high flow releases are proportional to the natural flow regime, including base flow contribution in the low flow season through regulated ground water use" (MoWR 2012: 4).

A systems approach to river basin management can be considered as an improved alternative,—often referred to as "Pareto Improvement" in economics. River basins are sensitive over space and time; any single intervention has implications for the system as a whole. Activity taking place in a part of the basin (e.g., disposal of wastewater, deforestation) will have impacts downstream. A vivid example of this was the cyanide spill in the River Tisza (a tributary of the Danube) from a mine in Romania in January 2000. The highly toxic chemical swept downstream through Hungary, devastating aquatic life along the course of the river and contaminating the drinking water of hundreds of thousands of people (WWF 2002). The other example is the construction of the Farakka barrage in 1975 on the lower Ganges in India. The idea of constructing this barrage was to divert water to resuscitate Kolkata port. However, over time excessive sedimentation in the barrage led to stream-flow depletion further downstream along the natural course of the Ganges, especially in the estuarine zones (Rudra 2004; Bandyopadhyay 2012a; Danda et al. 2011). There have been ecosystem losses in the form of mangrove depletion and other species loss, as also to livelihoods (Bandyopadhyay and Ghosh 2009; Bandyopadhyay 2012b; Mukherjee 2011).

While today's best practices in water resources planning entail integration of water quantity and quality management for both groundwater and surface water, there remains a need for a comprehensive understanding of how the natural environment and the resident population of a basin are impacted by various levels of interventions in the rivers or by adoption of new policies, land use as well as land and vegetation management. This is best done in a highly participative way, involving all the major stakeholder groups, and in a way that achieves a balance between the level of economic development and the consequent impact on the natural resource base of a river basin as agreed to by the stakeholders. This participatory and comprehensive approach is what is generally referred to as good integrated river basin management (IRBM).

11.6 Knowledge Gaps in Relation Between Water Resource Use and Food Production: IWRM, IRBM, and Low-Carbon Growth Pathways

Food production has traditionally been thought of as an increasing function of land and water. Therefore, the necessary condition for achieving food security is perceived in the form of supply-side interventions by bringing in more land under agriculture, and water resource development projects. Both these entail exploitation of nature, and interventions in the eco-hydrological cycles. While land-use change has critically affected livelihoods, the other critical knowledge gap between the developed and the developing nation is the divergence in the understanding of perceived relation between water and food. The delinking of water from growth, as is perceived in the understanding of IWRM, conforms to the principles and pathways to low-carbon growth, as will be discussed in Sect. 11.6.1.

Traditional food policies perceive food security as a positive function of water availability. Recent literature, however, refutes such a relation (e.g. Förare 2008; Ghosh and Khan 2012). Developed nations have been lately emphasizing demand management of water, and development of institutions such as markets to achieve efficient allocation. Considering their huge and harmful impact on ecosystems and consequently on livelihoods, large dams are being decommissioned in many parts of the developed world. It is being recognized that merely satisfying short-run agricultural needs without thinking of the sustainability of ecosystem services might be counterproductive for long-run food security, considering their linkages with the provisioning services of ecosystems. Various resource-saving practices, such as the System of Rice Intensification and newer irrigation techniques, are also being innovated. Various experiments have refuted the direct proportionality between water and food availability and this knowledge is increasingly being recognized in the policy frameworks of the US, EU, and many other parts of the developed world. Bandyopadhyay and Ghosh (2009) have highlighted the deficit of this knowledge in South Asia, and have emphasized on the need for creating the knowledge base on the water–food nexus.

While Falkenmark et al. (2007) infer that by 2050, food demand will roughly double; the demand for water allocations for agriculture will rise, as also demand for land. Though plausible options lie in increase in water use and expansion of agricultural lands, supply-side interventions such as water augmentation plans by building large dams, and bringing in more land under agriculture by cutting down forests, as also unsustainable use of fertilizers and pesticides, will have deleterious impacts on the ecosystem, threatening the very ecological foundation of food provisioning services in the long run. Therefore, there is no doubt that the trade-off between the short-term economic needs and the ecosystem sustainability (with its implications on long-term food security) exists.

Options therefore need to be sought in demand-management practices, and institutional reforms. From a very regional food security perspective, trade in "virtual water" (or agricultural imports) can indeed play a crucial role. Each option has its own implication for the nonagricultural ecosystem and the services they generate. However, for serving the longer term economic needs of food security, a more holistic perspective is needed. This will entail an integrated approach for managing land and water resources and ecosystems that acknowledges the multi-functionality of agro-ecosystems in supporting long-term food production.

Undoubtedly, long-term planning is subject to uncertainty. Most of the tools developed so far for dealing with trade-offs (spatial or temporal) involving ecosystem services work best when ecosystem behavior and responses to external stimuli are known and understood. However, ecosystems are hardly subject to such certainty. Outcomes are therefore unpredictable and difficult to control (Falkenmark et al. 2007). However, ad hoc decisions might prove counterproductive. Decisions related to tradeoffs under uncertain conditions should be based on a set of alternative scientifically informed arguments, considering the entire eco-hydrological cycle, and the trade-offs that may exist with the interventions into this cycle. As such, even institutional mechanisms like Payment for Ecosystem Services (PES) may be thought of. PES can work in agriculture, where ecosystem services are threatened and the opportunity costs for alternatives are not very high (Ottaviani 2011).

11.6.1 Low-Carbon Growth Pathways

While IWRM talks about an ecosystems approach to managing water and in the process talks of river basin as the planning unit, there is no doubt that this creates the right pathway for low-carbon growth. Various publications recognize this characteristic. In a recent publication, UNESCAP (2013) recognizes that adopting IWRM and treating river basin as the planning unit through IRBM creates opportunities for converting water resource constraints and threats on environmental security in various hydro-political relations into opportunities. More importantly, IWRM talks of keeping water in-stream thereby helping the ecosystem services many of which are provisioning and regulatory in nature. More interestingly, the food provisioning service, which has been discussed in this paper, helps in food production through natural processes rather than the energy consuming technology. Keeping water in-stream and non-conversion of forestland to agricultural lands augments regulating services such as climate regulation, carbon storage, and sequestration.

The critical aspect here is with the acknowledgement that water also has a supporting service. In its floodplain, water supports forest and biodiversity, as also providing important provision services. The forest biodiversity, in its turn, plays important provisioning, regulating, cultural and supporting services. Natural water recharge, water purification, etc. are a few of such services. Further, the role of water in the carbon sequestration process of the forests is a common public knowledge.

Ghosh (2009) lists the various ecosystem services provided by water, and talks of the importance of valuation of economic and ecosystem services to understand the trade-off arising out of prioritization of water use between the various sectors. Definitely, there is a critical trade-off for approaching water management through IWRM, and for approaching water management through the mechanism of the traditional supply intervention plans which cannot be the pathway for low-carbon growth.

In this context, let me reiterate what I have already argued in this paper as an important mechanism for adopting IWRM, i.e., use of economic instruments, which entails scarcity value-based pricing of water. This has also been argued by Ghosh and Rachuri (2011). The biggest problem in the developing world, and more so in India, is that irrigation water is subsidized to the extent of being almost zero. It is hardly recognized that this subsidy is a veiled tax on the ecosystem demand for water, which hampers the long-term food security. Subsidized electricity essentially adds fuel to this fire, and in many parts of India (e.g., the Punjab-Haryana belt with the prevalence of the rice-wheat cycle; the Cauvery River basin where from the 70s onwards, cultivation of a less water consuming *Ragi* was replaced by paddy), this led to depletion of groundwater. Hence, proper pricing of water should be thought of clearly to prevent the reprehensible wastage of water in agriculture. As argued by Ghosh and Rachuri (2011), for embracing the low-carbon growth path, there is a need for water prices to reflect the scarcity value of ecosystem services.

11.7 Concluding Remarks

Food security is an important policy concern in India. However, production mechanisms followed so far have been resource-intensive, and more so from the perspective of water resources. Policy documents have failed to embrace an integrated agricultural and water policy, as they fall under two different ministries. But the fact remains that agriculture consumes more than 85% of the total accessible water, and more so because of the process of subsidization. As argued in this paper, in its attempt to promote low-carbon growth, therefore, there is a need to promote less water-intensive agricultural practices in India. But, a micro-level attempt is not sufficient. There has to be a broader macro-policy mechanism that needs to be put in place. In this context, this paper attempted to highlight a few things.

Therefore, the following set of messages becomes important in the context of the sustainability of the ecosystem, water, and food nexus that needs to be embedded in IWRM. First, ecosystems are crucial for providing long-term food needs of the human society, and this needs to find explicit recognition in policy documents. Food security is not a linear function of water use, and there needs to be more emphasis on demand management of water rather than supply augmentation. Second, an integrated management approach is needed for land, water, and the ecosystems at the basin level to enhance the multiple benefits, and minimize the detrimental effects on the ecosystem services. Third, there is an urgent need to develop institutional and economic measures to prevent ecosystem degradation, and encourage changes in the practices of business-as-usual. Pricing of agricultural water is an important element here. Fourth, there is a need to develop less resource-intensive practices (e.g., System of Rice Intensification) for producing crops that have traditionally been high-resource-consuming. Fifth, policy documents need to explicitly recognize that the relation between water and food is not necessarily linear. Rather, irrigation development projects (like large dams) might even have detrimental impacts on food availability and livelihoods in the long run. Sixth, solutions to the problems of food security need not be sought in water supply-side management alone, but more emphasis needs to be placed on distributional and the demand-side aspects as well. This can help in a more integrated approach to water management, while considering the release of pressure on the ecosystems. Seventh, this also goes well with the emerging literature on environmental flows that ask the question of how much water the river needs. A river basin approach is needed to promote the needs of the ecosystem keeping in view the ecosystem livelihoods linkage. Eighth, IRBM needs to be taken up as the doctrine governing the management of water, and adequate institutional arrangements should be put in place to promote the paradigm in practice. These few messages become critical for water systems management in order to create a pathway for low-carbon growth.

References

Acreman M (2000) Background study for the World Commission of Dams, reported in World Commission on Dams (2000)

Alagh YK, Pangare G, Gujja B (eds) (2006) Interlinking of rivers in India: overview and Ken-Betwa link. Academic Foundation, New Delhi

Amarasinghe U (2012) The national river linking project of India: some contentious issues, www.iwmi.org/iwmi-tata/apm2012. Last viewed on 7 Apr 2015

Bandyopadhyay J (2004) Adoption of a new and holistic paradigm is a pre-condition for integrated water management in India in G. Saha (Ed) Water Secu Manage Water Resour (Kolkata: National Atlas and Thematic Mapping Organization)

Bandyopadhyay J (2009) Water, ecosystems and society: a confluence of disciplines. Sage, New Delhi

Bandyopadhyay J (2012) Water science in India: Hydrological Obscurantism. Econo Politi Week 47(16):45–47

Bandyopadhyay J, Ghosh N (2009) Holistic engineering and hydro-diplomacy in the Ganges-Brahmaputra-Meghna Basin. Econo Politi Week 44(45):50–60

Bandyopadhyay J, Perveen S (2008) The interlinking of Indian rivers: questions on the scientific, economic and environmental dimension of the project. In Mirza MQ, Ahmed AU, Ahmad QK (eds) Interlinking rivers in India: issues and concerns. Taylor and Francis, Abingdon pp 53–76

Barbier EB, Thompson JR (1998) The value of water: floodplain versus large-scale irrigation benefits in northern Nigeria, Ambio 27(6):434–40

Boelee E (Ed.) (2011) Ecosystems for Water and Food Security. UNEP and International Water Management Institute (IWMI), Nairobi and Colombo

CBD (1992) Convention on Biological Diversity 1992 (United Nations)

Chaturvedi V (2015) Cost of inaction on mitigating climate change. (This Volume)

Danda AA, Sriskanthan G, Ghosh A, Bandyopadhyay J, Hazra S (2011) Indian Sundarbans Delta: a vision. World Wide Fund for Nature-India, New Delhi

Falkenmark M (1997) Society's interaction with the water cycle: a conceptual framework for a more holistic Approach. Hydrol Sci 42:451–466

Falkenmark M (2003) Water Management and Ecosystems: living with Change. TEC Background Paper 9. Global Water Partnership, Stockholm

Falkenmark M, Gottschalk L, Lundqvist J, Wouters P (2004) Towards integrated catchment management: increasing the dialogue between scientists, policy-makers and stakeholders. Water Resour Dev 20(3):297–309

Falkenmark MM, Finlayson L, Gordon EM, Bennett T, Chiuta D, Coates N, Ghosh M, Gopalakrishnan RS, de Groot G, Jacks E, Kendy L, Oyebande M, Moore GD, Peterson JM, Portuguez K, Seesink RT, Wasson R (2007) Agriculture, water, and ecosystems: avoiding the costs too far. In: Molden D (ed) Water for food, water for life: a comprehensive assessment of water management in agriculture. Earthscan, London pp 234–277

Flessa KW (2004) Ecosystem services and the value of water in the Colorado River delta and Estuary, USA and Mexico: guidelines for mitigation and restoration. International seminar on restoration of damaged Lagoon environments, Matsue, Japan, pp 79–86

Förare J (ed) (2008) Water and food. The Swedish Research Council Formas, Stockholm

Foster AL (2003) Models for governing: opium and colonial policies in Southeast Asia, 1898–1910. In Go J, Foster AL (eds) The American colonial state in the Philippines: global perspectives. Duke University Press, Durham, N.C

Ghosh N (2008) A new look at integrated water resources management from the perspective of scarcity value of water resources. J Res Energy Develop 5(1):27–48

Ghosh N (2009) Economics of hostile hydropolitics over transboundary waters: scarcity values and interstate water conflicts in India and US. Saarbrucken VDM Verlag, Germany

Ghosh N (2012) Challenges to environmental security in the context of India-Bangladesh transboundary water relations: an agenda for research. Paper prepared for ISEE 2012. International Society for Ecological Economics, Rio De Janeiro

Ghosh N (2015) Challenges to environmental security in the context of India-Bangladesh Trans-boundary Water Relations. Decision 42(2). Special Issue on 'Managing Critical Resources, Food, Energy and Water

Ghosh N, Bandyopadhyay J (2009) A scarcity value based explanation of transboundary water disputes: the case of the Cauvery River Basin in India. Water Policy 11(2):141–167

Ghosh N, Khan SE (2012) Situation analysis on environmental security. A Bangladesh-India Initiative, IUCN. International Union for Conservation of Nature Ecosystems for Life, India

Ghosh N, Rachuri S (2011) Pricing the fluid mosaic: integrated Inclusive Valuation of water from the scarcity value perspective. India Infrastructure Report (3i Network and Oxford University Press)

Gilmartin D (1994) Scientific empire and imperial science: colonialism and irrigation technology in the Indus Basin. J Asian Stud 53(4):1127–1149

Gilmartin D (1995) Models of the hydraulic environment: colonial irrigation, state power and community in the Indus Basin. In: Arnold David, Guha Ram (eds) Nature, culture and imperialism: essays on the environmental history of South Asia. Oxford University Press, Delhi, pp 210–236

Gleick PH (1998) The World's Water 1998–1999: the biennial report on freshwater resources Washington DC: Island Press

Gleick PH (2000) The world's water: 2000–2001: the biennial report of freshwater resources. Island Press, Washington DC

Homer-Dixon TF (1994) Environmental Scarcities and Violent Conflict: Evidence from Cases. Int Secur 19(1):5–40

MA (2005) Ecosystems and Human Well-Being: Synthesis. Millennium Ecosystem Assessment. (Washington, D.C: Island Press)

Martinez-Alier J (2012) Environmental justice and economic degrowth: an alliance between two movements. Capital Nat Social 23(1):00001

Mattas C, Voudouris KS, Panagopoulos A (2014) Integrated groundwater resources management using the DPSIR approach in a GIS environment: a case study from the Gallikos river basin, North Greece. Water 6:1043–68

Molden D, Fraiture D (2004) Investing in water for food, ecosystems and livelihoods, Discussion draft, Comprehensive Assessment of Water Management in Agriculture, Stockholm

MoWR (2012) National water policy 2012. Ministry of Water Resources, Government of India

Mukherjee J (2011) No voice, no choice: riverine changes and human vulnerability in the 'chars' of Malda and Murshidabad". Occasional Paper No. 28. Institute of Development Studies Kolkata

Osemeobo GJ, Njovu F (2004) An evaluation of woodland utilization by smallholders in the Central and Copperbelt Provinces of Zambia. Int J ApplEcono Economet 12(2):219–236

Ottaviani D (2011) The Role of PES in Agriculture. In Payments for ecosystem services and food security. FAO, Rome

Richardson RB (2010) Ecosystem services and food security: economic perspectives on environmental sustainability. MSU International Development Working Paper 110, Department of Agricultural, Food, and Resource Economics, Michigan State University, December 2010

Rudra K (2004) The encroaching ganga and social conflicts: the case of West Bengal. Independent Broadcasting Associates, Littleton, MA

Sen AK (1981) Poverty and famines: an essay on entitlement and deprivation. Oxford University Press, Oxford

UNESCAP (2013) The Status of the Water-Food-Energy Nexus in Asia and the Pacific. Position paper commissioned by the United Nations Economic and Social Commission for Asia and the Pacific

Webb P, Coates J, Frongillo EA, Rogers BL, Swindale A, Bilinsky P (2006) Measuring household food insecurity: why it's so important and yet so difficult to do. J Nutr 136:1404S–1408S

WWF (2002) The ecological effects of mining spills in the Tisza river system in 2000. World Wide Fund for Nature

Part III
How can India Grow in a Low Carbon Way?

Chapter 12
Energy Security Options for India in the Context of Great Power Rivalry Emerging in the Indian Ocean

Subhomoy Bhattacharjee

Abstract The level of energy trade that swims across Indian Ocean is massive. While the ocean is far smaller than the Pacific and the Atlantic in its geographical spread, yet about 36 million barrels per day or about 40% of the global oil supply accounting for about 64% of oil trade use this stretch of water. Consequently, the Ocean has become the biggest challenge for maritime security and global sea trade. Securing energy security in this environment means ensuring that the energy lines are kept safe so that the onshore fuel dependent sectors can make long-term plans for investment. This would also depend on how well an economy builds up its navy and shipping power to address these challenges. This paper tries to look into the questions such as, HOW deep is the Indian commitment to this massive combination of opportunities and threats at her shores? How strong is the policy level and financial commitment to make inroads in the most significant geo-strategic change happening in the world economy? Are there elements of an active and integrated response taking shape?

12.1 Introduction[1]

Dokalam was not a stand off between India and China for the rights to control any mineral reserves, current or anticipated. What the flash point demonstrated though is that in the twenty first century, the rivalries between nations that begin to grow their economies are as strident as they were in the previous centuries. This has implications especially where the rivalry concerns the use of scarce energy resource. India and China have large and growing needs for energy resource either in the shape of oil, gas, or coal to feed their economy. The potential of renewable

[1]"This paper draws in some respect from a PolicyBrief written by the author for RIS and available on its website www.ris.org.in".

S. Bhattacharjee (✉)
Business Standard Newspaper, New Delhi, India
e-mail: s.bhattacharjee@ris.org.in

R. Kathuria et al. (eds.), *Low Carbon Pathways for Growth in India*, India Studies in Business and Economics, https://doi.org/10.1007/978-981-13-0905-2_12

energy to create a new paradigm of energy supply in this environment is a work in progress. It will take time, upwards of two decades to whittle the fundamental constraints of the energy matrix.

Nowhere is that constraint more visible than in the Indian Ocean. The level of energy trade that swims across the Ocean is massive. While the ocean is far smaller than the Pacific and the Atlantic in its geographical spread, yet about 36 million barrels per day or about 40% of the global oil supply accounting for about 64% of oil trade use this stretch of water.[2]

Most of the key energy suppliers of the world had straddled this shore line. The addition is the emergence of three key energy importers, China, India, and Japan all of whom demand unfettered rights to navigate this sea lane. Add the fast growing energy starved economies of Pakistan, Bangladesh, Thailand, and Vietnam in the mix, it is fair to say the Ocean is now crowded with competing actors. It has become even more so as the preponderant part of global trade in gas from Qatar and Iran is also dependent on this sea lane.

Consequently, the Ocean has become the biggest challenge for maritime security and global sea trade. Securing energy security in this environment means ensuring the energy lines are kept safe so that the onshore fuel dependent sectors can make long-term plans for investment. This would also depend on how well an economy builds up its navy and shipping power to address these challenges.

Simultaneously, the challenges are drawing in large investment options. The most prominent of these are the plans for the two largest competing economic zones of the world—the Belt and Road Initiative, anchored by China and the Asia Africa Growth Corridor, jointly anchored by Japan and India. The Dokalam flashpoint demonstrates that when the stakes are high unlike that of a mountain pass, there could be standoffs with attendant impact on energy security.

How deep is the Indian commitment to this massive combination of opportunities and threats at her shores? How strong is the policy level and financial commitment to make inroads in the most significant geo-strategic change happening in the world economy? Are there elements of an active and integrated response taking shape?

12.2 Sea of Discord

In less than 6 months of 2017, the world economy has woken up to three major news developments each of which has one common geographical chain. They underscore the importance of the Indian Ocean.

The first of these was the formal launch of the Belt and Road Initiative by China in May, 2017 with the Indian Ocean outreach component labeled as the Maritime

[2]Competition in the Indian Ocean: Eleanor Albert, May 2016, Council on Foreign Relations https://www.cfr.org/backgrounder/competition-indian-ocean

Silk Road Initiative. While it has several components, what drew most attention within the initiative was the China Pakistan Economic Corridor. One end of the corridor is Gwadar, the deep sea port being developed by Pakistan in its Baluchistan province with Chinese support estimated at $51.5 billion.[3] China because of its geography does not have an access to the Indian Ocean and so its co-development of the port on Arabian Sea has drawn attention. In June, Beijing has topped it by announcing it will set up its first military base overseas in Djibouti, in Africa. As Chinese naval ships sailed into the East African port, they will join US and French bases already positioned there. The expansion has of course been couched in diplomatic language to drive the host nations' "economic and social development". The language has not adequately answered the question why should a nation need an overseas military base that too in an economically busy channel, used by its rivals.

The second news development was the announcement of sanctions on Qatar by a host of Middle East states led by Saudi Arabia. The religio-terrorism angle has its salience but it is pivotal to note that each of the countries involved in the drama around Qatar depends on international trade in oil and natural gas. The two most busiest straits in the world, the Strait of Malacca[4] and the Strait of Hormuz,[5] between them account for shipping volume of about 36% of the total production of these commodities. In the crowded lanes, ships and consequently national tempers can collide.[6] Overall, the third largest ocean of Earth now accounts for close to two-thirds of the global seaborne trade in all commodities. No wonder President Trump and Prime Minister Modi stressed on expanded "maritime security cooperation", and also "determined to expand their engagements on *shared maritime objectives* and to explore new exercises".[7]

The third is the announcement by India and Mauritius in the month of May 2017, of plans to develop the island nation as a regional hydrocarbons trading hub. Mauritius, under its "Vision 2030", wants to develop itself as a "regional petroleum hub" to include both oil and gas.[8]

Each of these news emphasizes the vital role of this sea lane. This is not surprising. In the past two decades, as the growth engines of world economy have veered to the shores of Indian Ocean, it is inevitable that the role of Indian Ocean will become significant, much more eventually, than the ones played by the Pacific and the Atlantic Oceans. Consequently, the engagement with the Ocean has stirred up interest among all the littoral states and those beyond. The most visible

[3]https://defence.pk/pdf/threads/with-a-new-chinese-loan-cpec-is-now-worth-51-5bn.452582/.

[4]Malacca Strait transits grow 2% to record in 2015 http://bit.ly/2wwBiiA.

[5]240 foreign tankers call at Iranian oil terminal in wake of nuclear deal http://bit.ly/2wgCUNZ.

[6]USS John S. McCain collides with merchant ship near Strait of Malacca http://bit.ly/2fYbl5p.

[7]Joint Press Statement http://www.mea.gov.in/bilateral-documents.htm?dtl/28560/Joint+Statement++United+States+and+India+Prosperity+Through+Partnership.

[8]http://energy.economictimes.indiatimes.com/news/oil-and-gas/dharmendra-pradhan-meets-visiting-prime-minister-of-mauritius-mr-pravind-kumar-jugnauth/58866122.

manifestation of this level of interest is the rapid growth in the number of ports along the Ocean. They are currently 102 operational ports[9] but another 80 odd are under construction of which several are on the Indian coastline. However, one should not rush to conclusions. Of the 16 largest port regions of the world by tonnage, none are located on the Indian Ocean. The Atlantic Ocean trading system has more than 635 ports distributed on both shores.[10] Yet of the busiest ten shipping lanes of the world, the top four have Asia at one end.[11] Of them, only the Asia-North America route pass through Pacific, the others use Indian Ocean.

Within this vast canvas, what are the stakes for India to make a blue ocean outreach?

To answer the question, we consider the recent spate of attention on ports by India. Till recently, it was a back of the burner element of Indian economic policy. The other is the quest for an integrated energy policy.

12.3 Ports

It is important to emphasize at this juncture that there is no apparent reason why a stretch of water should play a spoiler for global trade, in this case energy. That it is doing so is however not surprising. There are two rationale intertwined here.

There is a lot of faith that the emerging investment in renewables will make the role of fossil fuels—a fossil, in the near future. Rapid improvements in the competitiveness of renewable energy imply that increases in this form of energy will "provide around half of the increase in global energy out to 2035". Still the fuel mix of oil and gas, together with coal, will remain the dominant sources of energy, even after two decades.[12] This means China and India will continue to pay a lot of attention to transport of those energy as their stock of these minerals (especially oil and gas) will have come down to single digits by then. So the importance of energy trade on these waters will rise instead of going down.

It follows consequently that positioning too many naval ships in close proximity of each other to shepherd the energy business has a neat way to generate friction. The Dokalam standoff this year and the ones before in previous years are examples of what happens when hostile military stand eyeball to eyeball. But as nations build up competing naval strengths to patrol their interests, it is progressively impossible to expect any of them to de-escalate tensions, more so as the geography of the Ocean lends itself of creation of asymmetric power play positions. The interplay of energy routes provides a possible fuse.

[9] List of ports http://ports.com/sea-browse/ocean/.

[10] ibid.

[11] Trade Routes http://www.worldshipping.org/about-the-industry/global-trade/trade-routes.

[12] BP Energy Outlook, 2017 edition.

India has the largest shoreline among all the nations which flank the Indian Ocean. As of now, it is impossible for India to replicate at a comparable scale the build-up of ports across the island nations that China has begun to build assiduously. To some extent, it is not even necessary because unlike Beijing, India sits on the Indian Ocean. Consequently, it can achieve a significant part of its objective by simply expanding its ports than it has invested in them in the past.

As of now, India has 12 major ports and about 200 minor ports.[13] A port-led strategy can have two objectives. It can secure the energy lines for disruption-free imports and that in turn, can consequently act as the catalyst for a long-term growth rate of 8% of annual GDP. According to an estimate,[14] the annual trade that passes the New Silk Route, from the Strait of Hormuz to the Strait of Malacca is estimated at US$18 trillion. Of this, the countries of the Gulf Cooperation Council ply seaborne trade of over $150 billion per year, meant for Indian and China. A disruption due to any reason in this volume can have a palpable impact on India's GDP. There has been no public estimate so far of the scale and potential of such possibility but clearly the government of India has become alive to the possibility. The emphasis on ports could be similar to the success of China, Korea, Japan, and Singapore, where their ports have led each of those countries' growth strategy.

To get a sense of the stakes involved, let us examine the economics of natural gas. Indian domestic gas production has decreased at a CAGR of around 10% from a peak of 46.04 BCM (billion cubic meters) as on FY11 to just 24.99 BCM in FY17. Consequently, imports have risen at a CAGR of 11% to make up the slack. In terms of comparison, India now imports 50% of its gas needs. It was only 22% in FY11.[15] Natural gas is expected to grow faster than oil or coal, helped by the rapid growth of liquefied natural gas increasing the accessibility of gas across the globe. The 2017 edition of BP World Outlook says while Asia will remain the largest destination for LNG, within the continent China and India will both increase their demand for LNG, helping the demand for the gas to grow faster than either oil or coal in each of these economies. Essentially, the Ocean acts as a giant pipeline for energy supply to both the Asian neighbors. Since there is little love lost between the two, needling the pipeline by acts of omission and of commission cannot be ruled out. At least after the punches exchanged this summer on the hills, it will be a naive commentator who does not give credence to similar possibilities on the sea.

This is the primer that illustrates why India has reached out to Sri Lanka and Maldives this summer.[16] Indian navy ships were drafted in fairy large numbers for flood relief and rescue operations off the coast of these countries. These measures tie in with the strategy that India has begun to explore to secure energy security for

[13]Indian Ports List… http://www.cybex.in/Indian-Ports-Data.aspx.

[14]http://english.alarabiya.net/en/views/news/world/2014/01/09/Why-all-eyes-should-be-on-the-Indian-Ocean.html.

[15]Robust RLNG Infra Imperative to Fuel Higher Usage of Natural Gas http://www.careratings.com/upload/NewsFiles/SplAnalysis/NaturalGas-RLNG.pdf.

[16]http://economictimes.indiatimes.com/news/politics-and-nation/delhi-pushes-indian-ocean-policy-amid-chinas-forays-in-the-region/articleshow/58888131.cms.

itself. Complementing the ports on the Indian soil is the offshore ports like the one on Mauritius and under exploration at Male, at Seychelles and the biggest of them all at Chabahar in Iran. Each of them has a role in acting as conduit for import of energy products including oil, gas, and coal. But all this costs money.

12.4 Financial Commitment

As outlined above, India has begun to respond by beefing up its investment on ports. As of 2017, foreign direct investment up to 100% under automatic route is already permitted in the port sector. This has created an ecosystem of public–private partnership projects in the major ports. As per government data, a total of 33 such projects with an investment of Rs. 17,817.96 crore are operational and another 20 with a projected investment of Rs. 22,362.67 crore are expected. These numbers are important as the corresponding level of investment by the government in ocean-related sectors has been minuscule as the chart shows.[17]

The big change in this respect is the launch of the Sagarmala Development Company (SDC) in July 2016 and the India Ports Global Private Limited, (IPGPL) a joint venture between Jawaharlal Nehru Port Trust and Kandla Port Trust incorporated in January 2015. The formation of the two in successive years and the scale of investments being planned for them show that GOI is moving towards an integrated response to the oceanic challenge. This response has been titled National Perspective Plan (NPL).

While SDC is meant for port investments within India, IPGPL is meant to do the same abroad. SDC however has a larger mandate than just ports. Other than setting up SPVs for domestic ports, it also aims to develop coastal economic zones. Both of them will raise funds from multilateral and bilateral agencies as debt and equity to finance the projects.

The NPL is ambitious. It seeks to push investments of close to Rs. 8 lakh crore in four streams. These are port modernization, connectivity enhancement, port linked industrialization, and coastal community development.[18]

The numbers need to be put in perspective, though. They are spread over a 20-year life span envisaging a yearly spend of Rs. 40,000 crore ($62 billion), not highly ambitious. But they demonstrate that

(1) GOI has abandoned public sector led model for development of ports
(2) The scale of investments is front loaded and expected to be harnessed mostly by 2025.

---

[17]Developing ports: Sagarmala Project http://www.makeinindia.com/article/-/v/developing-ports-sagarmala-project.

[18]Summary of projects under Sagarmala. PIB, 26 December, 2016 http://pibphoto.nic.in/documents/rlink/2016/dec/p2016122601.pdf.

As the data shows, of the projected doubling of the port capacity, the development of new ports (320 MMTPA) and of nonmajor ports (315 MMTPA) will be the growth drivers of this process. These are also the two areas where the bulk of the investment is anticipated. Between 2018 and 2025, GOI expects to receive a sum of Rs. 62,100 crore of investment from all quarters for these projects. On the basis of those inflows, GOI expects aggregate port tonnage to almost double for India from the current 1673 MMTPA to 3083 by FY25.

12.5 Integrated Energy Outcome

Despite the formulation of the Draft Energy Policy, GOI has still been reticent in bringing together the elements of a seaborne energy focus in play. While there has been a valuable recognition of the role of renewable energy, it has not been adequately supplemented with a focus on drawing together the overseas sources of supply of oil, gas, and coal. Yet just as we saw above from a serendipitous 22%, the share of imported gas has now reached close to 50%. This has implications for downstream industries. Investment in fuel demands a long gestation period. Indian industry has rarely seen any certainty of supply of any of these fuels to make plans stretching for decades worthwhile for industry to contemplate. Yet without such certainty, it is impossible for industry to create capacity in sectors that need fuel. The impact of making projections based on spot market imports has ruined the coal-based power sector companies.[19] After a brief respite, shortages of domestic coal have again come back to haunt the sector and make them explore avenues overseas. It is true that prices of coal are now at a historic low, but what riles manufacturers is the lack of knowledge of the future. GOI is squeamish about offering any long-term security in the crude oil markets too. As of now, a combination of appreciating INR and soft crude prices have kept import bills soft, but without offering any clarity on long-term purchases. The petroleum ministry estimates that GOI wants a 10% cut in the imports of refined products by 2022 and expects to raise the percentage to three-fourths of the demand from domestic sources to be met by 2040. But at the same time, it also admits "there has been gap between, innovation, industry and policy which was hampering the growth (of refining capacity)".[20]

Without bringing together these elements, an integrated response to the challenge of entering the Indian Ocean as an active player is still far out. Yet unless combined they will preclude India from operating from a position of strength in the waters. In this context, the recent government decision to dismantle the incipient formation of a ministry of energy is a retrograde step. As of now other than the ministry of shipping, there is no clearly defined agency to focus on all aspects of the

[19]India's Coal Story, Bhattacharjee, page 147; Sage Publications, India.

[20]http://bit.ly/2j7ZmDi.

contest for the waters. Ensuring the presence of government in the Ocean would need a remit that is larger than what the shipping ministry can provide. For instance, UAE (del Qatar) has offered to coordinate with India on two fronts-fill up India's strategic reserves of oil as well as ensure a committed supply of LNG via a pipeline. While the former has proceeded to some extent, the latter has yet to reach financial closure. The lack of an integrated response is the reason for the delay. Yet how India deals with this ally will also determine how India's own maritime position is addressed in future. Along with financial support, these sorts of responsibilities need an overarching decision-making framework within the GOI and the sooner it is addressed the better.

Chapter 13
Conclusion

Rajat Kathuria, Saon Ray and Kuntala Bandyopadhyay

Abstract This chapter draws conclusion to the main question asked in the book: how can a country grow in a low-carbon way? The book has examined this issue from the lens of the transport, industry and water sectors. Changes in lifestyle may be necessary to achieve energy efficiency in the household sector, with energy-efficient buildings in the commercial and residential sector. A move towards the use of energy efficient appliances will also be very important. All this will require access to finance and technology. Finally, innovative ways of thinking and the participation of every citizen of this country will be necessary to achieve the voluntary goals India has set for itself.

13.1 Conclusion

The world is likely to be very different in 2030, and predicting the future is hazardous. Predicting India's growth path and whether it is likely to be low carbon is similarly fraught uncertainty, with many scenarios possible. This book examines the way in which India can go the low-carbon way by 2030.

The concept of a low-carbon economy is defined by Jiang et al. (2010) as an economy that has a minimal output of greenhouse gases (GHG) into the biosphere, that is, an economic model based on small energy consumption, low environmental pollution, and low carbon emissions. It also entails changes in the country's industrial structure and people's conceptions. Among the key drivers of the low-carbon economy, Osborne (2010) identifies equipment and infrastructure that

R. Kathuria (✉) · S. Ray · K. Bandyopadhyay
Indian Council for Research on International Economic Relations,
New Delhi, Delhi, India
e-mail: rkathuria@icrier.res.in

S. Ray
e-mail: sray@icrier.res.in

K. Bandyopadhyay
e-mail: kbandyopadhyay@icrier.res.in

R. Kathuria et al. (eds.), *Low Carbon Pathways for Growth in India*, India Studies in Business and Economics, https://doi.org/10.1007/978-981-13-0905-2_13

enable energy efficiency or alternative energy production and use, leading to a reduction of carbon emissions, directly or indirectly (via such mechanisms as smart buildings, smart grids, renewable energy, biofuel vehicles, electric vehicle charging systems). The key question is: how can a country grow in a low-carbon way?

India's INDC has stated that it will reduce emissions intensity by 30–35%, by 2030 from the 2005 level. This will require moving away from coal, which is the largest primary energy source to renewables.[1] The deployment of wind, solar, biomass, and hydropower will contribute towards this goal. While coal may dominate power generation in the coming years, using cleaner coal and moving towards supercritical technology will help improve energy efficiency.

Energy efficiency measures need to be implemented in all sectors of the economy. The book has examined this issue from the lens of the transport, industry, and water sectors. However, energy efficiency in the agriculture and power sectors will also be critical. Changes in lifestyle may be necessary to achieve energy efficiency in the household sector, with energy efficient buildings in the commercial and residential sector. A move towards the use of energy efficient appliances will also be very important. All this will require access to finance and technology. Finally, innovative ways of thinking and the participation of every citizen of this country will be necessary to achieve the voluntary goals India has set for itself.

References

Jiang B, Sun Z, Liu M (2010) China's energy development strategy under the low carbon economy. Energy 35:4257–4264

Osborne TM (2010) Carbon capital: the political ecology of carbon forestry and development in Chiapas, Mexico. Ph.d. thesis Energy & Resources, UC Berkeley

[1]http://www4.unfccc.int/submissions/INDC/PublishedDocuments/India/1/INDIAINDCTOUNFCCC.pdf.

The manufacturer's authorised representative in the EU is Springer Nature Customer Service Centre GmbH, Europaplatz 3, 69115 Heidelberg, Germany. If you have any concerns regarding our products, please contact ProductSafety@springernature.com

Printed and bound by CPI Group (UK) Ltd, Croydon, CR0 4YY
25/09/2026
02222900-0002